Materials Technology in the Near-Term Energy Program

Report of

THE AD HOC COMMITTEE ON
CRITICAL MATERIALS TECHNOLOGY
IN THE ENERGY PROGRAM

TO THE SCIENCE ADVISER TO
THE PRESIDENT

COMMISSION ON SOCIOTECHNICAL SYSTEMS
NATIONAL RESEARCH COUNCIL

NATIONAL ACADEMY OF SCIENCE
WASHINGTON, D.C.   1974

NOTICE

The project which is the subject of this report was approved by the Governing Board of the National Research Council, acting in behalf of the National Academy of Sciences. Such approval reflects the Board's judgment that the project is of national importance and appropriate with respect to both the purposes and resources of the National Research Council.

The members of the committee selected to undertake this project and prepare this report were chosen for recognized scholarly competence and with due consideration for the balance of disciplines appropriate to the project. Responsibility for the detailed aspects of this report rests with that committee.

Each report issuing from a study committee of the National Research Council is reviewed by an independent group of qualified individuals according to procedures established and monitored by the Report Review Committee of the National Academy of Sciences. Distribution of the report is approved, by the President of the Academy, upon satisfactory completion of the review process.

This study, by the Commission on Sociotechnical Systems, was conducted under Contract No. NSF-C310, Task Order No. 297 between the National Science Foundation and the National Academy of Sciences.

Members of the National Research Council study groups serve as individuals contributing their personal knowledge and judgments and not as representatives of any organization in which they are employed or with which they may be associated.

The quantitative data published in this report are intended only to illustrate the scope and substance of information considered in the study and should not be used for any other purpose, such as in specifications or in design, unless so stated.

Requests for permission to reproduce this report in whole or in part should be addressed to the National Research Council-National Academy of Sciences.

For sale by the National Technical Information Service, Springfield, Virginia 22151

AD HOC COMMITTEE ON CRITICAL MATERIALS TECHNOLOGY
IN THE ENERGY PROGRAM

COMMISSION ON SOCIOTECHNICAL SYSTEMS

NATIONAL RESEARCH COUNCIL

Steering Committee

Harvey Brooks, Harvard University, *Chairman*
Seymour Blum, Mitre Corporation
Alan Chynoweth, Bell Laboratories
Richard Claassen, Sandia Laboratories

ad hoc Committee Members

Richard S. Claassen, Sandia Laboratories, *Chairman*
James H. Bechtold, Westinghouse Research Laboratories
Richard J. Charles, General Electric Research and
  Development Center
Arthur C. Damask, Queens College
Albert M. Hall, Battelle Memorial Institute
Franklin P. Huddle, Congressional Research Service,
  Library of Congress
Riki Kobayashi, Rice University
Robert I. Jaffee, Electric Power Research Institute
Elburt F. Osborn, The Carnegie Institute of Washington

Professional Staff

Chester W. Spencer, National Research Council
Donald G. Groves, National Research Council
Farlan Speer, National Research Council
W. Lee Garner, Sandia Laboratories

Liaison Representatives

S. Victor Radcliffe, National Science Foundation
Robb Thomson, Federal Energy Administration

The speakers, visiting experts, and staff who made this study possible are listed in the Appendix. The committee is indebted to the visiting experts for their invaluable knowledge and help provided. The opinions expressed in this report, however, remain the responsibility of the committee.

The chairman would like to acknowledge help received from Sandia Laboratories' staff on several topic areas in this study.

ABSTRACT

A brief study of the near-term energy program (1985) in the United States identifies areas where materials technology may play an important role. A methodology provides a semiquantitative measure of the influence which a given materials program might have on the energy supply/demand balance in 1985. The resulting sensitivity indices indicate that only a few program areas can have significant impact in this time frame. The most important programs from the materials point of view with respect to degree of impact are (1) pressure vessels in the nuclear power plants, (2) oil shale, (3) coal liquefaction, (4) fuel and materials recycling from municipal and agricultural waste, (5) coal gasification, (6) high temperature turbines, and (7) hot water geothermal. If all of these particular programs are supported, the materials program cost over the next decade would total $470 million.

Other programs considered are judged to have less impact in this time frame: solar heating, extractive metallurgy processing, fuel cells, U_{235} separation, and batteries and flywheels for energy storage.

Implementation of most of all these programs may be severely restricted by availability of certain material elements and supplies of steel plate and forgings.

Research areas common to many of the energy programs are briefly summarized.

PREFACE

In March of 1974 the Science and Technology Policy Office, in support of Dr. Stever, the Science Adviser to the President, approached the newly designated Commission on Sociotechnical Systems of the National Research Council (NRC) about the feasibility of organizing a short term study of critical materials research and engineering needs in connection with the implementation of Federal energy goals. More specifically, the desire was for a study which might be helpful in objectively identifying

- in a semiquantitative way those critical material developments that will determine the feasibility of the technologies being recommended to meet the nation's needs for energy over the short period of 1975-1985;

- materials work that will result in a reduction in the use of primary fuel or will allow a substitute of other fuel for oil or gas.

Although the activity was to be organized under the Commission on Sociotechnical Systems, the importance of the energy program and the breadth of the study requested required the involvement of several groups in the NRC. Therefore, I invited Dr. Richard Claassen, chairman of the Solid State Sciences Committee of the Assembly of Physical Sciences; Dr. Alan Chynoweth, chairman of Panel 4 of the Committee on Mineral Resources and the Environment (COMRATE), and Dr. S.L. Blum, chairman of the National Materials Advisory Board (NMAB) to meet with me as a steering committee to define the scope of the project and to recommend individuals for membership on an *ad hoc* committee which would conduct the study in the form of a two-week workshop to be held at Airlie House, Warrenton, Virginia, in July 1974, and to be headed by Dr. Claassen.

The charter was stated in the original letter of assignment as follows:

"The Science and Technology Policy Office, in addressing the national policy in energy self-sufficiency has concluded that it desires a study of priorities in materials research relevant to national energy goals... Criteria will be evolved in the study wherein it will be possible to identify the materials related issues in the national energy system of 1985. Criteria will also be evolved whereby it will be possible to measure the impact of the different materials-related issues on the national energy system. The final report will include suggestions concerning future studies that might be addressed to materials-related issues involving the national energy system beyond 1985. It is expected that the recommendations of the study will be incorporated into decisions of government agencies responsible for the disposition of resources directed to materials research and development in the energy area..."

The *ad hoc* committee was established with Richard Claassen as its chairman and Don Groves of the National Research Council professional staff as the committee officer. The present report is a summary of the findings of the workshop, based on the knowledge of the participants, and on testimony from experts in each of the general areas of energy-related research covered in the report. The focus of the study was on the short term, that is, on the identification of research or development that could have some potential impact on energy supply or demand during the period prior to 1985. And in fact, the results of the study are presented in terms of their potential impact on the total energy supply-demand picture in that year.

Relative to the lead times for research, the year 1985 is very near term, especially when one takes into account the fact that a substantial investment to embody the results of research must be made before any appreciable impact on energy supply or demand can be realized. Thus most of the research problems identified relate to energy research and development programs which are already quite well along towards the demonstration plant or device stage. It would be misleading to infer from this report that the particular research areas singled out for discussion are the only ones that should have high priority in a national energy research and development program. Many programs started today could have an important impact on energy only after 1985.

Obviously, in the short time available, it is not to be expected that new ideas or directions will be turned up. Rather the report is an attempt to pull together in one place the state of the art as regards materials in a considerable variety of areas of energy demand and supply. Even in such a short term research program, however, it must be remembered that unexpected discoveries or previously unidentified problems will show up as the program proceeds. Any report such as this can be no more than a snapshot of the existing state of knowledge; it must not be treated as a rigid blueprint for the future, but rather as a tentative assessment which should be periodically re-examined as new knowledge becomes available, perhaps even on a yearly basis.

I must express my gratitude to the members of the task force who organized themselves so effectively on such short notice, as well as to the numerous consultants who contributed to the workshop. The enthusiasm with which this task was embarked upon by all participants was testimony to the high importance which the technical community attaches to the nation's energy program.

                                        Harvey Brooks, Chairman
                                        Commission on Sociotechnical
                                            Systems

                                        September 4, 1974

CONTENTS

	Page
EXECUTIVE SUMMARY	1
INTRODUCTION	11
METHODOLOGY	13

MATERIALS ASSESSMENTS IN PRINCIPAL ENERGY PROGRAMS

    Coal Gasification 25
    Coal Liquefaction 32
    Oil Shale 37
    High Temperature Gas Turbines 39
    Critical Elements 43

MATERIALS ASSESSMENTS IN OTHER ENERGY PROGRAMS

    Nuclear 47
    Energy Conservation Through Materials Management 51
       Metals Extraction and Processing 57
       Fuel From Waste 62
    Geothermal 67
    Solar Energy 70
    Energy Storage 73
    Fuel Cells 76
    Isotopic Separation of Uranium$_{235}$ 79

OTHER ISSUES 81

RECURRENT RESEARCH THEMES 84

FUTURE STUDIES 86

APPENDICES

    Study Participants 91
    Coal Gasification 93
    Coal Liquefaction 102
    Automotive Gas Turbine Engines 103
    Critical Elements 105
    Batteries 115
    Flywheels 118

REFERENCES 121

FIGURES

Figure Number		Page
1	Energy supply-demand pattern in the United States in 1985 (Brookhaven Base Case).	14
2	Schematic of energy sources, conversion and flow.	15
3	Energy consumption for production of materials and by principal consumer use.	16
4	Cumulative nominal heat generating capacity of all nuclear power plants in the U.S. (actual and projected) and the nominal heat content of the high BTU coal gas from plants projected, as a function of year.	26
5	Cumulative capital investment in the U.S. in nuclear power plants and in high BTU coal gasification plants versus year.	27
6	Generalized liquefaction process flow diagram.	34
7	Typical load curves for electric power generation for one week.	40
8	Specific fuel consumption versus percentage horsepower for several turbine engines.	103
9	Forecast of average miles per gallon for automobiles using piston engines versus introduction of ceramic turbine engines in 1990.	104

TABLES

Table Number		Page
1	The sensitivity of the near-term energy program to materials effort in several programs.	18
2	Coal gasification processes heat content and primary uses.	28
3	Energy consumption in the United States by use sector in 1972.	52
4	Energy consumption in the United States by end use in 1973 (estimated).	53
5	Industrial sector energy consumption in Quads.	54
6	Sixteen industries (or products) which accounted for an estimated 50% of the total industrial consumption in 1968.	56
7	Energy required to produce three primary metals.	58
8	Yearly organic waste generation in U.S.	62
9	Components of municipal refuse (dry basis)	64
10	Electric power costs for geothermal, nuclear, hydropower and coal plants.	68
11	Battery characteristics.	74
12	Selected coal gasification processes.	95
13	Plant investments required for ceramic gas turbine auto engines and cumulative oil savings.	103
14	Critical elements uses and sources.	112
15	Non-energy uses of critical elements.	114
16	Technical objectives for batteries used for load leveling and vehicles.	115

TABLES (continued)

Table Number		Page
17	Comparison of today's secondary batteries.	116
18	Electrochemical cells under development for bulk energy storage.	117

EXECUTIVE SUMMARY OF THE AD HOC COMMITTEE ON CRITICAL MATERIALS TECHNOLOGY IN THE ENERGY PROGRAM

INTRODUCTION

In the summer of 1974, an *ad hoc* committee of the National Research Council addressed materials aspects of the nation's near-term energy program (1975-1985) to assist the Federal Energy Administration in the formulation of a detailed blueprint for implementation of Project Independence. The urgency of the schedule required that attention be focused upon short-range contributions of an engineering or application nature, eliminating from detailed consideration such long-term energy programs as fusion, breeder reactors, *in situ* processes for oil and coal, and others. In recognition of long-term needs, however, the committee would endorse a broad, aggressive, research effort directed toward longer-range solutions where materials sciences could make a contribution.

METHODOLOGY

A simple methodology is presented which systematizes the selection of near term problems requiring investigation. It also provides a rough measure of the probable result of allocating material development resources to particular areas within the energy program. A sensitivity index is developed which estimates:

1. The effect of individual materials programs upon the energy supply-demand situation in 1985, measured in Quads* produced or saved,

2. The capital requirements in billion dollars per Quad,

3. The cost per million BTU,

4. The committee's level of confidence in the successful accomplishment of the materials portion of a program,

5. Materials development costs, and

6. A measure of the leverage that a given materials program investment can have on an energy program in the near term.

---

* A Quad (quadrillion BTU) = 10^{15} BTU. Note: For purposes of this report, we have considered the terms, Quad and Q to be synonymous.

Several principal energy programs are assessed in considerable detail; other energy programs which appear less certain by 1985 are addressed, but with relatively less emphasis. These two groups of programs are summarized below:

PRINCIPAL ENERGY PROGRAMS

Coal Gasification: Projections of energy contributions and capital investment requirements for coal gasification in 1980 and 1985 are compared with similar quantities for nuclear power industry. If the projections are to be verified, the investment in the next 10 years in coal gasification must be at least as great as that which has been made to date in the nuclear power program. Commercially proven processes exist for making medium BTU coal gas. The goal is to develop large scale plants for high BTU gas which will be economically competitive. A substantial materials effort will be required to develop a mature coal gasification technology that will exceed the capability of the Lurgi process. Objectives of the newer processes include the use of a wider variety of coals, lower initial capital costs, and improved process efficiency through higher reactor pressure.

Materials problems in coal gasification arise from the increased pressures and temperatures required in the improved processes, the larger scale of the equipment, and the severe service demands upon pressure vessels, pumps, valves, heat exchangers, and piping and power conversion components at the combustion interface. Physical and chemical effects include hot erosion and corrosion arising from impurities in the coal and from hot massive solid by-products of the process. The welding and inspection of field-erected pressure vessels will require special techniques. A materials test facility similar to that used in the nation's nuclear program would provide simulated service conditions for evaluating materials and components. Materials science support is required. The *ad hoc* committee estimates that the total materials program costs will be about $320 million over the next decade.

The most important constraints in achieving coal gasification objectives are likely to be shortages of: heavy steel plate, large steel forgings and castings, chromium, skilled engineers and tradesmen, and capable engineering and construction organizations.

Coal Liquefaction: Coal liquefaction is an extension of current industrial processes for the catalytic hydrodesulfurization of heavy, high-sulfur residual fuel oils. This program should not place a heavy demand on materials development. One significant problem with coal liquefaction, however, is erosion-corrosion effects when materials are exposed to sulfidation environments. A recent Ohio State University workshop on the problem concluded that the erosion phenomenon is poorly understood in the coal conversion processes. Funding on erosion-corrosion of about one-half to one million dollars annually is required to support the coal liquefaction program. Shortages in manpower, fabrication capability, and critical items (such as stainless and chromium-molybdenum reactor steels) also threaten this program. The committee emphasizes the need for acquiring, analyzing and distributing information relative to (1) failures and their analyses, (2) nonproprietary unique construction details, and (3) quality control and quality assurance programs.

Oil Shale: By 1985, production of oil from oil shale will face the basic constraint of competitive world petroleum prices, as well as domestic pollution standards, water availability, leasing practices by the government, and socio-economic problems attending a large new industry in sparsely inhabited regions.

Aside from the expected long delivery lead times for large steel retorts and heavy equipment, no major problems associated with materials were identified.

High Temperature Gas Turbines: There are two developing areas related to high temperature gas turbines that could affect 1985 energy programs: turbines for use in electric generating plants for intermediate duty and turbines for use in automobiles. Three approaches are underway to achieve higher operating temperatures and efficiencies: (1) improvements through metallurgical development to create advanced alloys, protective coatings and composites; (2) application of new high-temperature ceramics for stationary and rotating turbine blades; and (3) cooling of turbine blades by circulation of a fluid through blade ducts. If the turbines are designed to operate on feed from gasified coal rather than fuel oil, materials problems will arise, since the particulates and chemical impurities entrained in the coal gas will corrode turbine blades. Materials development costs are estimated at $70 million.

Application of ceramic gas turbines to automobiles will require low cost design for mass production. Introduction of significant numbers of automobile turbines is not likely before 1985, but the potential savings in the succeeding years suggest that the solution of remaining materials problems might be worth the projected $25 million materials development investment that would be required to perfect the automotive turbine.

Critical Elements: The committee identifies nine chemical elements and minerals in an approximate order of short-range criticalness for energy programs: manganese, chromium, fluorspar, nickel, cobalt, aluminum, tungsten, platinum and copper. Criteria for evaluation are: essentiality for the energy program, extent of reliance upon imports, potential adequacy of substitutes, and vulnerability of imports to concerted control of price or flow

Attention is focused on specialty requirements of the steel industry. Fluorspar is critical since it is essential to both steel and aluminum production. For each of the nine materials, two types of action are considered: (1) supply-demand corrections, and (2) "quick payoff" research and development for the purpose of new alloy development.

Payoff for research and development within the 1985 time frame could come from new alloy development, preparation of handbooks about high performance alloys with low critical material content, and construction of demonstration plants capable of improving the critical materials balance. Program costs are expected to be about $95-125 million over the next decade.

OTHER ENERGY PROGRAMS

The second tier of programs which were subjected to less extensive discussion by the committee includes the following:

Nuclear: In 1985 nuclear power reactors are expected to generate an order of magnitude more electrical power over 1974. The committee acknowledges that a large, competent, diverse community is already involved with development and design and with the identification of important materials problems. Therefore, only one problem area is considered: excessive downtime as a result of failures in the steam subsystems of nuclear plants. These failures are similar to those which occur in fossil fuel plants, but when they occur in nuclear plants, the larger capacity and the greater concern for safety usually cause a longer downtime, affecting both power quantities and costs. Stress corrosion in the system is now avoided by keeping the water within a narrow alkaline range by means of chemical additives. Even so, local flow patterns and scale deposits still cause some corrosion. Although design changes could bring about further improvements, more laboratory work is needed on understanding the nature of stress corrosion, in particular within the chemical and physical environments actually experienced. Research on condensers, in particular, could lead to elimination of defects which now cause transfer of contaminants from the outside cooling water to the steam system.

Further knowledge is needed concerning welding and inspection techniques to assure continuance of existing high standards and safety. Pressure vessel delivery dates (by whatever fabrication process) are dependent upon manufacturer capacity and availability of appropriate metals. If domestic capacity is not enlarged, it is estimated that half the pressure vessels needed by 1985 will have to be made overseas. Possible lack of control over the quality of scrap introduced in overseas manufacture raises a potential problem of long-time interaction of radiation-produced defects in steels containing impurities. Elimination of this uncertainty will require a systematic study of the radiation damage to steel with different combinations and concentrations of impurities. Activities to calculate reliability of pressure vessels will require the participation of universities, government laboratories and industrial firms. Such interaction will produce a better understanding of that reliability, a wider agreement on estimates of reliability, and a higher degree of confidence in design calculations. An effort involving 20 principal investigators for a period of 5 years (about $10 million costs) would yield a significant improvement in this important area.

Conservation: Proper management of materials processing and fabrication can conserve energy in significant amounts. The committee finds it difficult to generalize about measures which specific industries might adopt, but calls attention to the opportunities, to the needs for quick response, sustained motivation and effort, and systematic engineering study. The potential savings are 10 Quads in 1980 and perhaps as much as 20 Quads by 1985. The greatest opportunities for industrial energy savings through materials management lie in the areas of process steam, direct heat, electric drive, feed stocks and electrolytic processes. Detailed accurate information about energy use by standard industrial classification, particularly about the energy consumed in the recovery of ore and manufacturing processes is badly needed.

About 13% of the nation's energy consumption is devoted to the production of material, and a large part of that energy lies unsalvaged in dumps. Some energy is recoverable through recycling of energy-intensive metals such as aluminum and copper and by burning combustible wastes to make direct heat or by converting waste to fuels. Designs which allow the use of materials which are less energy-intensive could also assist in reducing energy demands.

Extraction and Processing: Eight percent of the country's energy is used for mining and beneficiation of ores, primary smelting, refining, and primary fabrication of steel, aluminum and copper. Rising costs of energy and ever leaner ore bodies will require more heavy equipment for metal mining technologies-- placing another burden upon steel producers and manufacturers. Beneficiation costs could be reduced through automated control of the size reduction and separation necessary to liberate the mineral values from the waste rock. Leaching processes could be modified or developed to avoid grinding and comminution. *In situ* technologies could be developed to further increase energy savings.

Because natural gas and fuel oil are rising in price, smelters will be shifting to coal for fuel; present systems already possess (or will soon be required to possess) facilities for removal of sulfur oxides from stack gases. Recovery of manganese nodules from theocean floor will require support of public funds in this expensive and risky undertaking. The Hall process now used for the production of aluminum is energy intensive. A new chloride process will reduce energy consumption over the Hall process by 30%, but will probably be introduced only in new plants. Improvements in the Hall process itself (at a 15% savings) are also possible through improved heat balances, bath chemistry and electrode materials development.

In ferrous metals, savings could be effected with an extension of continuous casting process which would allow retention of most of the first heat in the product. Attractive features are offset, however, by capital investment and plant inflexibility.

The committee believes that a program concerned with the near-term energy problem should also provide for university research in this field as well as the training of new mining and extractive metallurgical engineers.

Fuel from Waste: The energy content of the total waste produced each year in this country represents about 12% of the total energy consumed, but only a portion can be recovered (half the urban, 1/30th the agricultural and 1/11th the forest wastes). (Respective energy content is 1.9, 0.2, and 0.8 Quad annually.) Urban waste recovery requires separation processes to improve efficiency of combustion and recovery of valuable metals. Separation processes are many and present varying materials problems (shredder faces, pipelines and feeder mechanisms, for example, are subject to erosion). In the incineration process, boiler tubes and refractories are corroded by hydrochloric acids generated by pyrolysis of chlorine-containing wastes, particularly PVC plastics, and by ferrous oxides in the slags which are generated. A materials program of $20 million is needed.

Animal waste recovery is still in its infancy but represents a potential recovery of 26 million barrels of oil annually. The preferred technology of pyrolizing the waste to oil or gas by chemical change through heat is equally applicable to other forms of cellulosic waste such as paper, wood and sewage sludge. Since the process involves current boiler pressures and temperatures, no materials problems are expected. A low-pressure pyrolytic process, involving high temperature and the presence of ammonium sulfate, may give rise to materials problems, however.

Wood waste produced in lumbering processes (hog fuel) is frequently used in lumber mills to generate process steam and could be used by power companies to generate peak load electricity, but the relatively small amount of raw material recoverable does not warrant a large research effort into current erosion problems with grates and pneumatic feed lines.

Geothermal: Sources of energy are now confined to the dry steam geysers area of California where electric power costs compare favorably with those of nuclear, hydropower and coal-based plants. This well-developed technology requires no materials developments in the near term. If liquid-dominated geothermal fields are to be as extensively exploited as steam-dominated fields, materials problems become significant. The typical brine of a hot water field produces erosion, corrosion and deposition problems with the well head, steam pipes, separators and turbines, blades, nozzles and shaft seals. One way to avoid the salinity problem is to "flash" the water (suddenly reduce the pressure so that part of the superheated water vaporizes instantly) but as much as one-third of the heat is lost. Another way is to neutralize the effects of dissolved oxygen in the hot water by addition of sodium sulfite. Until new corrosion-resistant materials are developed (e.g., certain ceramics, tantalum, plastics, etc.), aluminum, chrome 10% stainless steel and cured resins will be the material candidates for hot water systems. A materials program of $15 million is needed.

Solar: Solar energy could be exploited successfully if materials could be designed into low-cost hardware for such high-use applications as water heating and space heating and cooling for residences. Although solar energy is free, clean and widely distributed; available at those times of the day and year when electric power demand is at peak; and could be designed to be modular and mass producible, it is also intermittent (requiring redundant systems), diffuse (requiring large arrays of collectors), and nondirectional (requiring tracking devices). Several existing technologies permit its adaption to new construction, but costs depend upon geographical location, fraction of load supplied by solar, and other factors. Substantial cost reduction could come from the large potential market and from new materials for collectors that are efficient and mass producible. Better corrosion-resistant selective absorbers especially need to be developed, either through electro-deposition or spray-on/paint-on techniques. Program costs are estimated at $20 million over the next decade for development of such materials and systems.

Energy Storage: Two energy storage devices, batteries and flywheels, will involve considerable materials efforts if they are to contribute to energy systems in the next decade. Rechargeable batteries can have special use in load leveling in the utility industry and for short-haul electric vehicles, if improvements in materials can extend lifecycles and reduce high capital costs. Sodium-sulfur and zinc-chloride cells are expected to have later-term impact after lead-acid cells have reached maximum development. Flywheel development will require materials with very high strength-to-density ratios--most probably fiber composite materials. Load leveling applications will not occur before 1985, but vehicle propulsion applications for trolley cars and buses may see limited adoption by that date. Near-term battery development will cost about $70 million over the next decade; flywheels will require materials support of about $15 million over that period.

Fuel Cells: Fuel cells offer advantages of silent, low-pollution, unattended operation; high efficiency at partial as well as full loads; and economy in modular units, allowing for distributed siting. For special need applications, where cost is not a primary consideration, fuel cells have performed with considerable success. They have some disadvantages: they require fuel of high hydrogen content and they are not economically competitive with other electric generators. Developments are underway with new systems, including acid systems, molten carbonate systems, base systems, and solid oxide electrolyte systems. The two most promising present materials problems which must be solved: The phosphoric acid system requires an expensive platinum catalyst whose effective life is about one year (80% of the platinum is recoverable). Fabrication procedures must be improved, and scaling to full production plants must be accomplished. In molten carbonate fuel cells, the electrolyte must operate at high temperature (600°C). Life times of the nickel or nickel-base electrodes need to be improved. A modest research investment ($3 million over 5 years) in dispersed metal electrochemical catalysts might have a large payoff in fuel cell technology.

Isotopic Separation of U_{235}: Processes other than gaseous diffusion have been studied to effect the isotopic separation of $Uranium_{235}$ for nuclear power reactors: centrifuge separation, whose ultimate limit of performance lies in the material strength of the rotating components; and laser isotope separation. Both processes could operate at about 10% of the gaseous diffusion energy expenditure, but further materials problems cannot be identified until the process of choice is defined.

OTHER ISSUES

The committee concludes that the materials community can solve the problems identified if given the challenge and opportunity. Thus, materials technology does not appear to be the limiting factor in the energy program. The serious near-term problem rather appears to be lead times for delivery of structural materials for large-scale plants for nuclear, coal gasification/liquefaction, and oil shale facilities, and their components. Major demands will be for low alloy and stainless steels and nickel-base super alloys. Objectives may not be reached for Project Independence unless steel plate-making capacity is doubled promptly. Forgings and castings may also face demand/supply deficiencies. An information system needs to be developed to determine industry capacities, anticipated demands, and supply situations so that large-scale computer modeling can properly deal with the tradeoffs involved. The committee also foresees a shortage of scientists and engineers trained and experienced in materials research and development and of skilled tradesmen and craftsmen in heavy construction industry.

RECURRENT RESEARCH THEMES.

Several recurrent themes are identified: The combination of erosion and corrosion, particularly at elevated temperatures, is a serious problem in nearly all the programs assessed. Higher reliability of design and operation are being demanded in many programs as a result of increased plant size and capital investments, higher pressures and temperatures, increasing public awareness, and closer supervision by government agencies. Better programs of testing and monitoring and quality assurance will be required. The importance of steel and catalysis technologies pervades the assessments. Catalysis will play a fundamental role in coal liquefaction, coal gasification, fuel cells and in some types of recovery of fuel from wastes.

FUTURE STUDIES

The modest near-term leverage upon the energy program available to the materials community is evident from the total of 17.2 Quads of energy projected to be produced or saved by 1985. Materials efforts over the longer term could produce larger dividends. It is essential to identify materials research programs which will have to start now to have down-stream effect. Some in particular hold special promise and present special problems: controlled thermonuclear fusion, direct conversion of solar energy to electricity, *in situ* oil shale and coal conversion processes, magnetohydrodynamic conversion, and breeder reactors. The methodology developed here could be helpful in setting problems in perspective and could be invaluable in highlighting the materials areas which must receive attention and resources, as well as environmental impact considerations.

APPENDICES AND REFERENCES

In the Appendices of this report, the committee presents further technical details of coal gasification technology and materials problems; areas of coal liquefaction research requiring assessment; automotive gas turbine efficiencies and economics; possible actions or potential solutions and research and development options for critical elements; and battery and flywheel technologies. A list of selected basic references is also provided.

INTRODUCTION

The Federal Energy Administration's Blueprint to detail the implementation of Project Independence will require the balancing of many parameters within several complex systems to determine the best way of minimizing dependence on foreign fuel sources. Since the Blueprint plan is to be reported in early fall, 1974, this study afforded the materials community an excellent opportunity to provide inputs to that larger decision-making process. The press of time has meant, however, that only a first portion of the overall problem could be addressed this summer. The scope of this study, therefore, is restricted to the near-term portion of Project Independence Blueprint. As an inevitable consequence, this criterion focused attention on short-range contributions of an engineering or application nature, and eliminated from consideration a number of energy programs which could have great importance in the long term and some of which are of extreme interest to the materials community.

Programs Not Considered

Fusion Power

   Magnetic Confinement
   Laser
   Electron Beam

Geothermal from Hot Rock

In situ Oil Shale

In situ Coal Gasification

Breeder Reactors

Solar, Other than Heating

Magnetohydrodynamic Conversion

Although the committee emphasized near-term solutions of an application nature, the membership and invited participants WOULD STRONGLY ENDORSE A BROAD, AGGRESSIVE, RESEARCH EFFORT AIMED AT THE LONGER-RANGE SOLUTIONS THAT MATERIALS SCIENCES COULD CONTRIBUTE TO THE NATION'S ENERGY PROGRAM.

To all of us, the need is obvious for a further comprehensive study of the materials problems in the energy program, with principal attention on the intermediate and long-range periods of Project Independence Blueprint. The dominant position of energy in our economy makes it mandatory that each technical community identify the contributions it can make to the national goals in the energy program. Comments on future studies will be found at the end of this report.

The entire spectrum of energy supply and consumption is continually influenced by materials developments, the great majority of which are of an evolutionary nature such as the increased working temperature of a superalloy or the extended life of a modified catalyst. In total, the many small improvements effect large results; but for this short study the committee sought representative examples where the material aspects played a key role in making significant changes. In an initial screen, the study participants selected for investigation those technologies which individually might produce, process, or conserve a minimum of 0.1 Quad annually (about 0.1% of the nation's total energy consumption projected for 1985 or 50,000 barrels of oil per day equivalent). Although some approaches were determined to fall short of this level, they are included in this report*.

The first section of the report provides the baseline for projections to 1985 and develops the method by which the relative importance of materials programs to the energy program was determined. A table there displays committee best judgment on a number of programs and culminates in a list of sensitivity indexes which provide a measure of relative benefits which might be expected from the expenditure of resources in different materials programs related to the energy program.

In selecting the particular energy programs for review and in scheduling time in the workshop, four areas stood out as requiring primary emphasis. They were: liquid fuels (including coal- and oil shale-derived), coal gasification, high temperature turbines and critical elements. These four areas are described first, followed by the problems on which, for various reasons, less time was spent by the committee.

Although the committee concentrated on materials technology rather than materials supply, it was continually reminded that lack of items such as plate steel may be the limiting factor in the energy program. The need for detailed forecasting of structural materials is described in the section on "Other Issues."

A number of broad research areas are common to several energy programs; these are briefly summarized. The final section of the report emphasizes the importance of material studies aimed at longer-range solutions. The Appendix contains additional details.

---

* Throughout the report, the unit of energy is a Quad = (quadrillion BTU) = 10^{15} BTU. This unit is gaining currency, is well defined and is of convenient size; e.g., the 1973 U.S. consumption was 76 Quads.

METHODOLOGY

At the outset of the study, a methodology was established to aid in systemizing the process of selection of problems for investigation. The methodology also provides a semiquantitative measure of the probable result of allocating material development resources to a given area of the energy program. The *ad hoc* committee study represented a special type of sub-optimization exercise, where the objective was to identify material problems relating to the near-term energy program (given as 1985).

One effective method to display the overall results is to generate a sensitivity index which attempts to estimate the effect that a particular materials program could have on the energy supply-demand situation in 1985, while all program factors other than the materials program are held constant. However, it is recognized that this is somewhat an artificial measure, since material programs alone cannot be meaningfully separated from the central development projects to which they contribute. On the other hand, it did seem most appropriate for members of the materials community to attempt to focus on their own areas of expertise.

The aforementioned sensitivity index is developed through a sequence of steps:

1. A baseline of the energy supply-demand situation in 1985 is accepted as given (see Figure 1).

2. The supply and distribution portion of Figure 1 is organized schematically in Figure 2. The consumption portion is displayed in Figure 3.

3. One program is selected and the impact on the energy supply-demand in 1985 is estimated.

4. Estimates are made of the capital investment required to produce (or save) each Quad in 1985.

5. The market price per million BTU is estimated.

6. The level of confidence that the materials development portion of the program will be successful in that time scale is judged by the committee.

7. An estimate is made of the material development costs over the next decade.

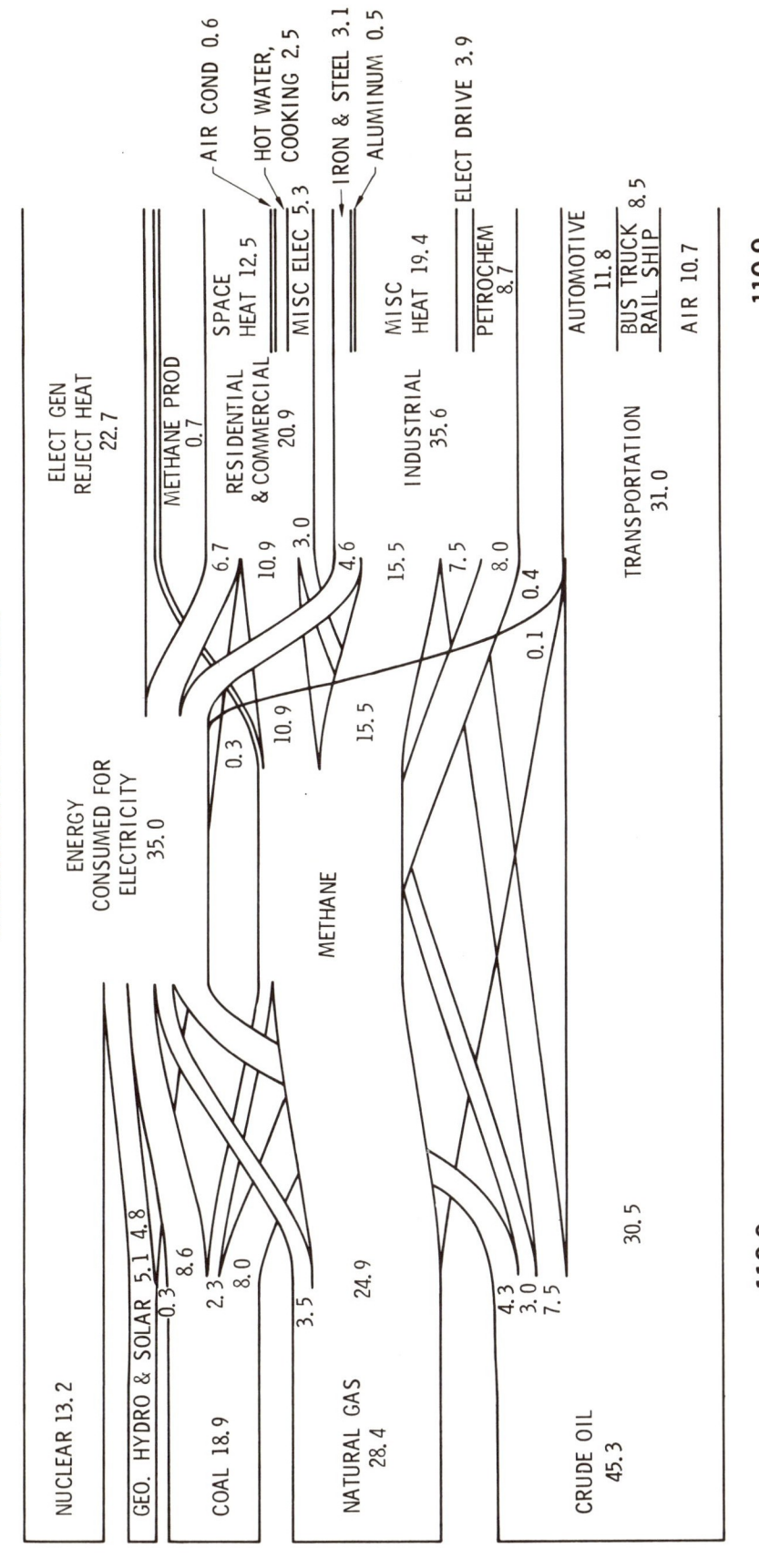

Figure 1. The energy supply-demand pattern in the United States in 1985 as projected by energy flow in Quads. (Brookhaven Base Case) This projection assumes considerable conservation. The nuclear portion of 13.2 Quads is equivalent to about 250 GWe plant capacity.

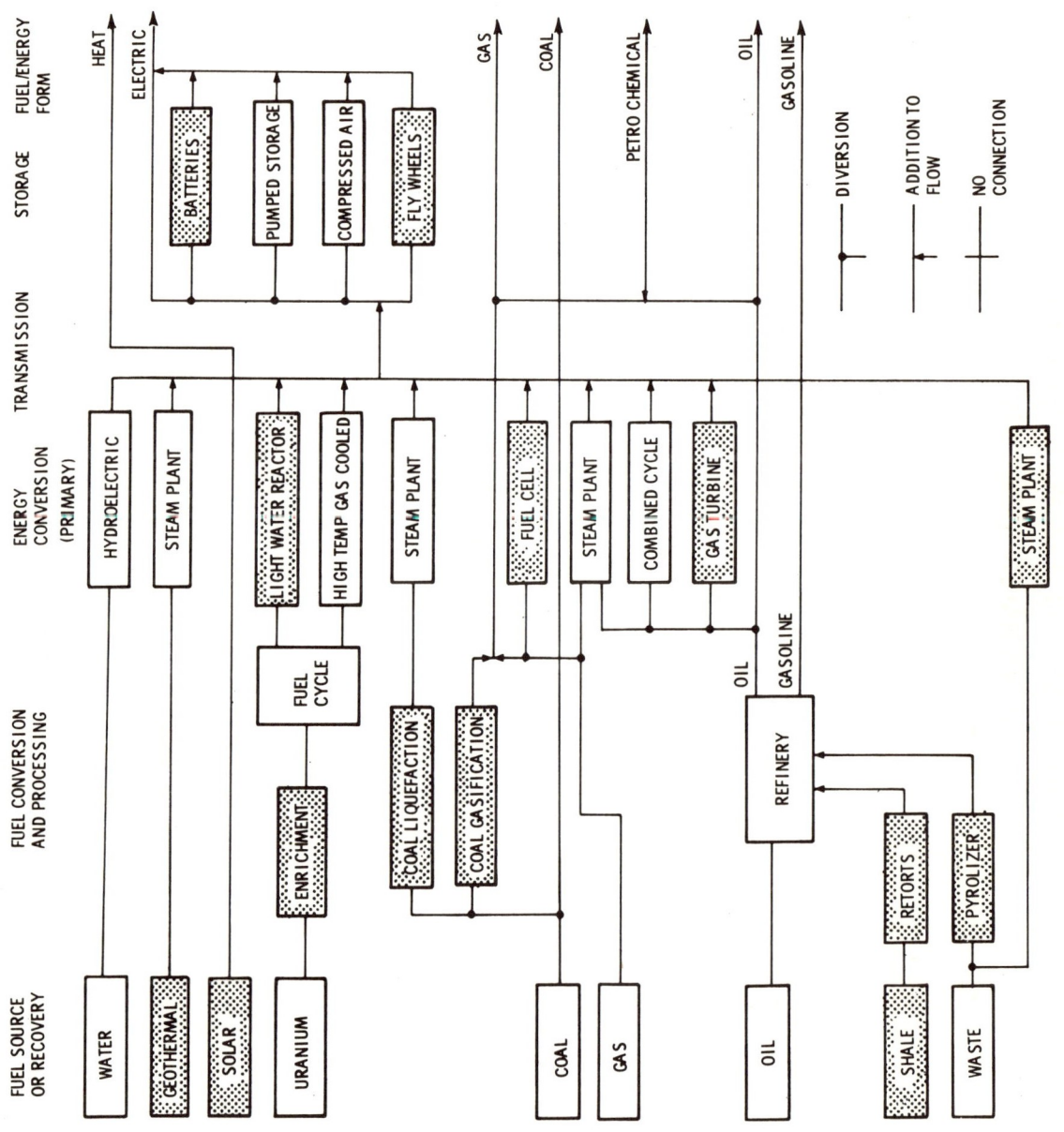

Figure 2. A schematic representation of energy sources, conversion and flow. Investigated in this study were materials aspects of the topics in the shaded boxes.

	FOR MATERIAL PRODUCTION	BY CONSUMER USE
Residential and Commercial		Space Heating Air Conditioning Water Heating and Cooking Miscellaneous Electric
Industrial	Iron and Steel Copper Aluminum Petroleum Refining Paper and Paper Board Petrochemical Feedstock Cement Ammonia Glass	Process Steam Electric Drive Fabrication Food Processing
Transportation		Air Automobile Bus, Truck Rail Ship

Figure 3. Energy consumption for production of materials and by principal consumer use, for three areas. The boxes are shaded to indicate that the listings are representative rather than complete.

8. A simplified scale is used for each of the above factors so that a sensitivity index can be computed by simple arithmetic.

A. Projection of Energy Supply-Demand

We have used a projection of data for 1985 (furnished by permission of the Associated Universities Inc. project at Brookhaven National Laboratory in advance of publication). A visual display of the data in a format developed by the Joint Committee on Atomic Energy is shown in Figure 1. The conclusions of our study would be little affected had we used another supply-demand projection such as an interpolation of those developed by the Joint Committee on Atomic Energy for 1980 and 1990. A strong energy conservation program might lead to a significantly different projection.

B. The Energy Matrix

The supply and distribution portion of Figure 1 can be organized schematically as displayed in Figure 2. Each box in this diagram is a potential area for investigation of programs in which materials efforts are required to make the projections in Figure 1 come true or which could modify Figure 1 in the desirable direction of reducing overall consumption of energy or in substituting another fuel form for natural gas or petroleum. Some boxes in Figure 2 are sensitive to materials, while others are far more dependent on other factors. Similarly, Figure 3 displays the major categories in consumption.

C. Program Impact on Energy

The real question for the near-term program is whether a given development project will contribute to the production or saving of energy in any significant way by 1985. For each program, we have attempted to estimate that impact measured by Quads produced or saved in 1985 (see Table 1). In coal gasification, for example, Table 1 projects 2.3 Quads compared with today's negligible amount. For high temperature turbines we estimate that 0.05 Quad could be saved in 1985 compared with use of more conventional technology for electric generators.

A scale of 1-5 is used to represent the energy factors, with 1 the lower limit at <0.1 Quad and 5 the upper limit at >10 Quads.

D. Capital Investment

Since the capital requirements for new energy-producing methods in this country will be so great, it is desirable to emphasize programs which minimize the capital

PROGRAM	DIFFERENTIAL ENERGY ADDED OR CONSERVED IN 1985 ①		CAPITAL COST ②		BTU COST ③		CONFIDENCE ④	VALUE $\frac{① \times ④}{② \times ③}$	MATERIALS PROGRAM COST ⑥		MATERIAL SENSITIVITY INDEX VALUE/COST
	Quads	Scale	B$/Quad	Scale	$/MBTU	Scale	Scale		M$	Scale	
COAL GASIFICATION	2.3	3	4.4	3	1.2	2	4	2.0	320	5	0.4
COAL LIQUEFACTION	0.5	2	2.6	2	1.1	2	5	2.5	10	2	1.3
OIL SHALE	1.1	3	5.2	4	1.2	2	5	1.9	MIN	1	1.9
STATIONARY TURBINES	0.05	1	5.0	4		3*	4	0.3	70	2	0.2
AUTOMOTIVE TURBINES	0.08	1	2.0	2		3*	4	0.7	25	2	0.3
NUCLEAR	12.0	5	2.0	2	0.7	1	5	12.5	10	2	6.3
FUEL FROM WASTE	0.8	2	8.3	4	0.4	1	4	2.0	20	2	1.0
GEOTHERMAL	0.3	2	2.4	2	0.5	1	2	2.0	15	2	1.0
SOLAR	0.1	2	50.0	5	6.0	5	4	0.3	20	2	0.2
BATTERIES	<0.1	1	14.0		1.2	3*	3	0.2	70	3	0.07
FUEL CELLS	<0.1	1	2.0	2		2	1	0.3	3	1	0.3
EXTRACTIVE METALLURGY	<0.1	1									
FLYWHEELS	<0.1	1							15		
U₂₃₅ SEPARATION	UNKNOWN										

*Neutral scale value

Footnotes.

	COLUMN 1		COLUMN 2		COLUMN 3		COLUMN 6	
Scale	Quads	Scale	Cap. Costs B$/Quad	Scale	Unit Costs $/MBTU	Scale	M$ Over 10 Years	
1	<0.1	1	<2	1	<1	1	<10	
2	0.1-1	2	2-3	2	1-2	2	10-50	
3	1-5	3	3-5	3	2-3	3	50-100	
4	5-10	4	5-10	4	3-4	4	100-200	
5	>10	5	>10	5	>4	5	>200	

Table 1. The sensitivity of the near-term energy program to materials effort in several programs. The quantities for each column are taken from the text (except for confidence, which is presented only in this table). The quantities are converted to a dimensionless scale of 1-5 as indicated in the footnotes. The value of program is estimated by multiplying the scales of columns 1 and 4 and dividing by 2 and 3. The sensitivity index is the value divided by the scale number for materials program cost. Note that this table deals only with materials program costs, not total costs of programs. Indexes for a later period would be different; and of course, the program list would expand to include such programs as breeder reactors, fusion, solar conversion, and others.

investment required to produce a given amount of energy per year. The units used are billion dollars per Quad. A B$/Quad is equal to about $2100 a barrel per day or $87 per kilowatt, assuming a 100% load factor. The capital investment factor is represented on a scale of 1-5 with the lower end at 1 representing <2 B$/Quad and 5 representing >10 B$/Quad.

E. Cost per Unit of Energy

The cost per unit of energy is an important element in considering the desirability of a given program. The market price 10 years from now for a process yet to be developed is a tenuous item at best. In addition, the proponents may have a tendency to develop figures which indicate competitiveness with present technology. Nevertheless, we present best estimates and a scale running from 1 for <1 $/MBTU to 5 for >4 $/MBTU.

F. Level of Confidence

The committee does not feel qualified to judge the overall probability of success for the various energy projects, but it does feel it appropriate to express degrees of confidence in the success of the material portion of the program. The level of confidence is indicated by a scale from 1 for low to 5 for very high.

G. Materials Development Costs

To the extent possible, costs for materials research and development, including the support for design and construction of operating systems, are identified. There is a reluctance to give quantitative estimates of the resources required for the material portion of a program extending over the next decade. A few of the reasons for this reluctance are:

- inadequate time was available for the committee to develop a thorough and complete material program for each project, and appropriate resources were not available to cost out such a program;

- the detail of inquiry varied widely between programs;

- estimates made with today's knowledge may be overturned by unforeseen developments in a project.

- forecasts made now may be misused in future years to justify irrational program decisions.

Nevertheless, the committee recognizes the necessity for program planners without a materials background to have some way of measuring the resources required in this area. Consequently, material development program costs have been estimated with the strong caveat that the only meaning is in comparing relative efforts. Any decision on project or program funding must be made on the basis of specific and detailed program descriptions which are far beyond the capability of a short-term, *ad hoc* committee. The estimate of materials development costs over the decade 1975-85 is represented by a scale from 1 for <10 M$ to 5 for >200 M$.

H. <u>Sensitivity Index</u>

High energy contribution or conservation and high program confidence are desirable, whereas low capital investment and low unit cost are preferred, so a program value can be estimated by:

$$\text{Value} = \frac{\text{scale 1} \times \text{scale 4}}{\text{scale 2} \times \text{scale 3}}$$

Dividing the resulting value by the scale for materials program cost yields a measure of the leverage a given materials program investment can have on the near-term energy program, or the

$$\text{Sensitivity Index} = \frac{\text{Value}}{\text{Cost}}$$

The various factors are discussed in the text (other than confidence). A summary is provided in Table 1. The scales chosen are, of course, almost entirely subjective. Some readers may prefer to develop their own scale factors to make comparison between programs.

Note carefully that these sensitivity indexes were developed for the <u>near-term</u>. A low index here does not necessarily imply <u>a low index</u> in the context of the next time frame.

The selection of those programs to be accelerated in the near future is an exceedingly complex matter involving economics, manpower limitations, legal and social questions, institutional problems and other factors. The Federal Energy Administration is addressing these difficult and interrelated issues. In this report we avoid recommendations which imply a judgment of the overall question, but instead provide an estimate of the resources required of the materials community to reach a certain goal, should that be adopted.

MATERIALS ASSESSMENTS IN

PRINCIPAL ENERGY PROGRAMS

Coal Gasification

From Figure 1 the contribution from gasified coal in 1985 is 2.3 Quads or 25 plants of 250 million standard cubic feet capacity per day. A substantial materials effort will be required.

PERSPECTIVE

It is instructive to view the scale of coal gasification in two ways: first, a consideration of the energy to be contributed per year, and second, the magnitude of the capital investment required. Nuclear power program data provide a convenient reference.

Figure 4 is a plot of nuclear capacity installed in BTU's per year versus the year. Actual numbers are displayed through 1973, projected numbers through 1979. For coal gasification, the projections for 1980 and 1985 are given.

Figure 5 is a similar portrayal of the cumulative capital investment against the year, based on 1973 prices. For the nuclear plants only the cost of the nuclear portion is shown. For the coal gasification plants, the investment required for extraction and delivery of the coal is not included.

These two graphs demonstrate that by 1985, the projected magnitude of the coal gasification program will be comparable to that of the nuclear program today. One can, of course, argue that these two programs are not directly comparable. Although nuclear energy was an entirely new field in 1964, most of the basic knowledge and principles in the nuclear field were well established. Today there are serious new problems in the coal gasification area which are sophisticated, subtle, and complex even though coal gasification is an extension of long established areas of technology.

In 1964, we knew how to make nuclear power plants; the problem was in making them economically competitive. In 1974, we have the processes to make coal gasification plants; the goal for the next decade is to produce them on a large scale, yet competitive economically.

In 1964, the budget for the Metallurgy in Materials Program in the Division of Research of the Atomic Energy Commission was $20.2 million. Using the Gross National Product (GNP) price deflater to adjust for inflation equates to the equivalent of about $28 million in 1974. In addition, other materials efforts directly funded by reactor programs in government and private industry were much larger than the AEC direct material research program mentioned above.

Even discounting this comparison, the fact remains that a substantial materials effort will be required in connection with the coal conversion programs.

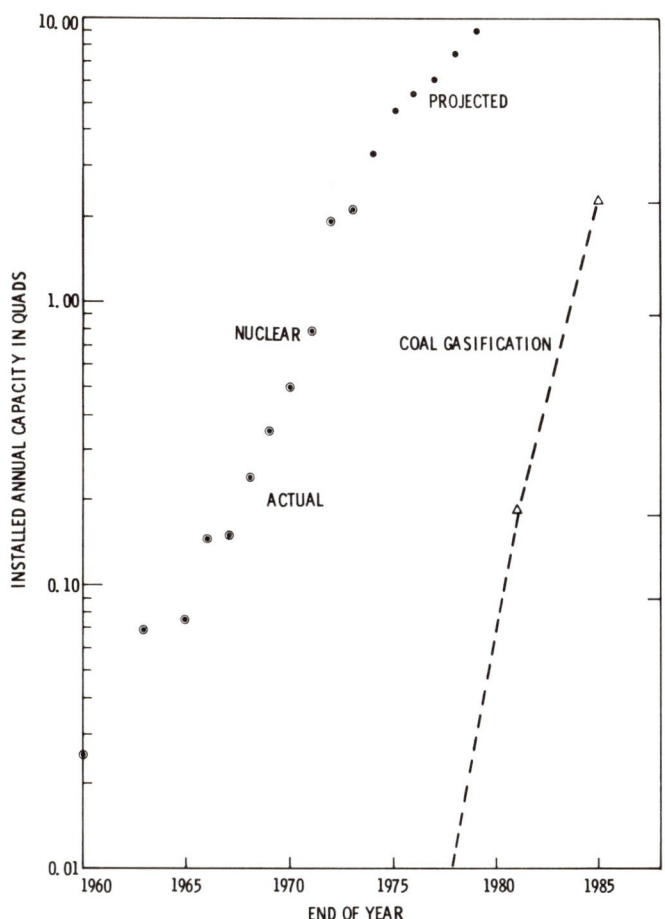

Figure 4. *The cumulative nominal heat generating capacity of all nuclear power plants in the U. S. (actual and projected) and the nominal heat content of the high BTU coal gas from plants projected to be in operation as a function of year. The nuclear figures assume 10,000 BTU/kWh; the gas contains 1000 BTU per standard cubic foot.*

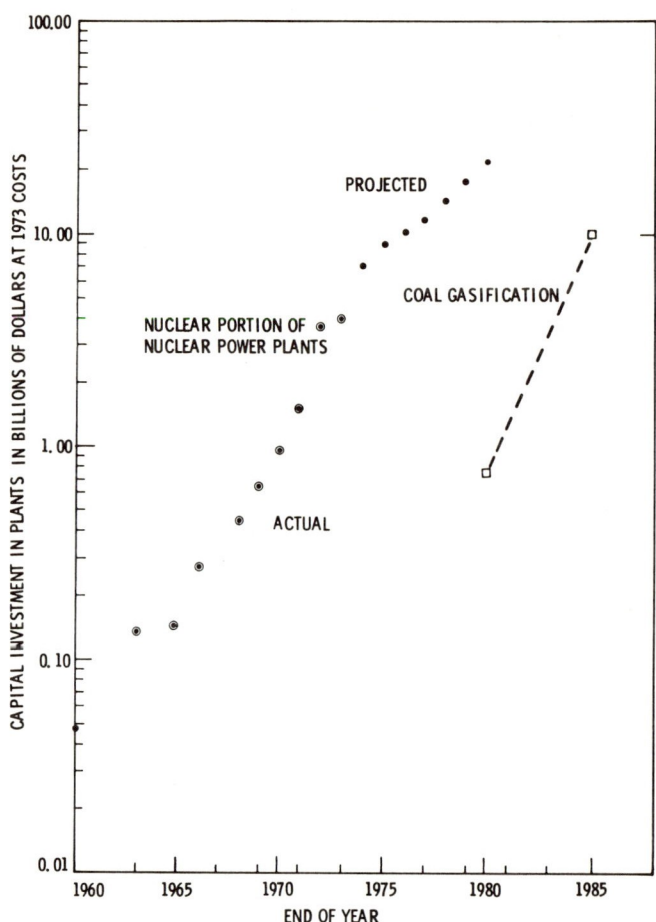

Figure 5. *The cumulative capital investment (in 1973 dollars) in the U. S. in nuclear power plants and in high BTU coal gasification plants versus year. The nuclear figures are based on a cost of $167 per kilowatt electrical for the nuclear or heat generating portion and 10,000 BTU per kWh. Comparable cost for the entire electric generating plant is $440 per kilowatt. The coal gasification figures are based on a 1973 estimate of $390 million for a 250 million SCF per day plant.*

TECHNOLOGY

The technology is available in the U.S. to convert coal into clean low or medium BTU gas (see Table 2) for industrial use and electric power generation. Conversion of coal to produce clean high BTU methane gas (Table 2) to replace natural gas for distribution to industry, commerce and the residential consumer is made by methanation of medium BTU gas, a process not yet demonstrated commercially. This gasification technology differs considerably from that for coal liquefaction because, while liquefaction entails modification of molecular structure, gasification requires the complete disassembly of the structure of coal. To produce medium and high BTU gas most economically will require high process temperatures where structural materials capabilities are severely pressed.

$\Delta Q = 2.3$

TABLE 2: Coal Gasification Processes Heat Content and Primary Uses.

Level	Heat Content (BTU per Std. Cubic Foot)	Primary Use
Low BTU	100-200	Electric generation process heat
Medium BTU	200-500	Feed stock for: methanation liquefaction
High BTU	1000	Replace natural gas in conventional systems

PROGRAM OBJECTIVES

Probably the most mature coal gasification technology is the Lurgi process. A Lurgi system producing medium BTU gas has been operating in South Africa for 20 years, and a Lurgi plant to produce 250 million standard cubic feet per day of high BTU gas is now under construction in northwestern New Mexico, with others being planned. The objectives of newer processes that are now in pilot or demonstration plant stage in the U.S. are:

- to permit use of a wider variety of coals than can be handled by the Lurgi process;

- to lower initial capital costs and subsequent operating budgets through construction of larger units; and

- to improve process efficiency by increase in reactor pressure.

The net effect of improvements arising from these objectives is estimated to be a 19% reduction in the unit cost of gas below comparable Lurgi costs.

PROBLEMS

Several processes under active development in this country at each of the three BTU levels are listed in Table 10 in the Appendix. Since most of the materials problems are common to many of the processes, they are discussed in their generic context.

The large scale of the required equipment, combined with the factors of high temperature and pressure, make the construction and operation of advanced gasification plants a technically challenging undertaking. In spite of this, present-day materials will be the basis for the construction of the first-generation plants, but these materials must first undergo extensive and detailed evaluation.

These coal gasification processes impose severe service demands upon major components. Pressure vessels, pumps, valves, heat exchangers and piping and power conversion components at the combustion interface are especially critical. In addition to the severe mechanical and thermal demands of the high-pressure, high-temperature conversion processes, there are additional critical chemical factors which arise in the refining of the coal. Sulfur, alkali, metals, steam, chlorides, and other impurities must be dealt with under a broad range of operating conditions.   MATERIALS PROBLEM

Problems are further compounded in most of the processes by hot erosion of the interiors of primary components. Fly ash, char, molten slag, and other solids must be transferred efficiently or entrained in rapidly flowing gases or liquids, or handled as hot massive solids.

In applying these materials, serious fabrication problems will arise that must be attacked at three levels: engineering in the field, evaluation in a test facility, and supporting materials science in the laboratory. Two problem areas require special attention:

1. The erosion, or the combination of erosion and corrosion, of valves and controlled-rate feeding devices, and

2. The welding and inspecting of field-erected pressure vessels which may be as great as 25 feet in diameter, 10-12 inches thick, and 250 feet high.

MATERIALS PROBLEM

A concerted attack on these two problems will require an effort of about $2 million per year until the problems are considered to be satisfactorily contained.

Other important problems include the unknown long-term reliability of pressure vessel shells, durable equipment for removing entrained solid particles from the gas, the durability of refractory-material liners, the life of catalysts, and the erosion/corrosion resistance of the critical components of power turbines fed by the product gas. Attention to these problems will require an additional effort of about $3 million per year.

MATERIALS PROBLEM

Besides these, there will be the usual variety of unanticipated problems with materials and components which typify the introduction of any new large-scale process system design and operation. Handling such problems will require an additional funding of about $5 million per year.

The attack on all these problems must be constituted as a well-coordinated coal gasification materials engineering research program.

MATERIALS TEST FACILITY

Effective resolution of the problems described, as well as improvement of materials for use in second-generation and subsequent plants calls for a separate test facility where materials and components can be evaluated under simulated service conditions. The value of such a facility has been amply demonstrated in the nation's nuclear power program. The coal gasification material test facility will require construction funding estimated at $100 million over 5 years and subsequent operation will require a further $7 million per year.

EFFORT REQUIRED

Economic production of gas in nationally significant quantities within the time frame set forth in Project Independence will require certain decisive actions:

1. Start the materials test, evaluation and fabrication programs immediately;

2. Establish and maintain close coordination between the design and materials inputs at all stages of the design and construction of equipment and plants;

3. Analyze material and component failures promptly and thoroughly, and report the results to all concerned agencies and individuals; and

4. Institute a supporting materials science program in such critical areas as erosion and corrosion.

The level of effort for the supporting materials science program is estimated to require $5 million per year. Like the other programs mentioned here, it should continue until the materials problems in coal gasification systems are adequately contained. Thus, the various funding requirements would result in a total expenditure of $22 million per year, plus $100 million expended over five years to construct the test facility.

Finally, there are added factors constraining the coal gasification program that were outside the scope of the committee's charge. Those of particular concern are listed below to lend additional perspective. Possible shortages are expected in:

- the availability of heavy steel plates;

- special large steel forgings and steel castings;

- the virtually indispensable alloying element chromium;

- the required number and mix of engineers and skilled tradesmen; and

- the needed number of engineering and construction organizations with the requisite capabilities.

This brief discussion is elaborated upon in the Appendix.

Coal Liquefaction

In 1985 coal liquefaction could contribute 0.5 Quad (from five 50,000 barrel-a-day units). Transfer of technology from the petroleum industry will be required as well as some materials development.

INTRODUCTION

The addition of hydrogen to coal in a suitable catalytic process can simultaneously remove undesirable sulfur and provide a fuel with the handling convenience of petroleum products. Addition of as much as 9% hydrogen results in conversion of coal to methane, ethane, propane, butane and gasoline. Hydrogenation is very expensive, however, by adding as little as 2-3% hydrogen a heavy liquid similar to Number 6 fuel oil can be produced. This makes an excellent feed for boilers designed for petroleum products and can be economically transported by pipeline where the generating station is substantially distant from the mine. For the near-term, then, coal liquefaction means, in effect, heavy fuel oil.

$\Delta Q = 0.5$

Coal liquefaction is not a new process, but is a continuation of the research begun some 50 years ago and applied by the Germans during World War II. Of the new technologies considered for U.S. energy production, coal liquefaction appears to be the one for which this nation is best prepared. Although it represents a considerable undertaking, much of the technology needed can be readily transferred from the petroleum industry's processes used to catalytically hydrodesulfurize heavy, high-sulfur, residual fuel oils. Most of the processes for the liquefaction of coal are similar to the hydrodesulfurization processes.

Following extensive research and development work in the United States, the first of the residual fuel oil hydrodesulfurization plants was made operational in 1968, and at least ten have been placed in operation since then, with an aggregate capacity of about 400,000 barrels per day. Acceptable performance of these plants is being routinely achieved on heavy Middle Eastern residual fuel oils in which sulfur content is being reduced from 4% down to the range of 0.1 to 1.0%. Although these plants were constructed primarily in Japan, the technology was provided by United States firms. Satisfactory performance is attested by (1) many companies have purchased a second plant and (2) on-stream time has been on the order of 90% (including time required to replace catalyst material and perform scheduled maintenance).

SIGNIFICANT PROBLEM AREA AND MAJOR EFFORT REQUIRED

One significant exception to the parallel between hydrodesulfurization of heavy, high-sulfur fuel oils and the liquefaction of coal is the considerably larger number of erosion problems encountered in coal liquefaction. For example, several of the pilot plants (such as H-Oil, COED and Cresap) have experienced major degradation of materials formerly assumed to be erosion-resistant. Moreover, the most frequently used erosion-resistant engineering materials (stellite and cobalt-bonded tungsten carbide) are not resistant to corrosion in sulfidation environments. Consequently, these materials are marginal for use in coal liquefaction processes. **MATERIALS PROBLEM**

At an April 1974 workshop held at Ohio State University on coal conversion processes it was concluded that, in the environments anticipated in coal conversion processes, erosion was a poorly understood phenomenon from the standpoints of theory, engineering approaches, testing standards, and specifications for erosion-resistant materials. Therefore, achievement of the goals for coal liquefaction will require a level of effort directed to these problems commensurate with their importance to the successful development of coal liquefaction. An appropriate effort would require one-half to one million dollars per year, until the problems are satisfactorily solved.

THE PROCESS

A typical process flow diagram for coal liquefaction is presented in Figure 6. The point of the diagram is to emphasize that it is virtually the same as those of the catalytic hydrodesulfurization processes for residual fuel oils. The presently proposed coal liquefaction processes also have in common with hydrodesulfurization the same requirements for materials of construction and very nearly the same operating conditions. Pressures of 2000-4000 psig and temperatures of 800-850°F are typical of both these processes. The steel alloy pressure vessels can be in direct contact with the process substances under these conditions.

Although much of the technology of the petroleum industry is directly transferable to the coal liquefaction program, it is very complex and can be undertaken best by those with demonstrated performance capability. As with most new developments, certain improvements will evolve in the normal course of

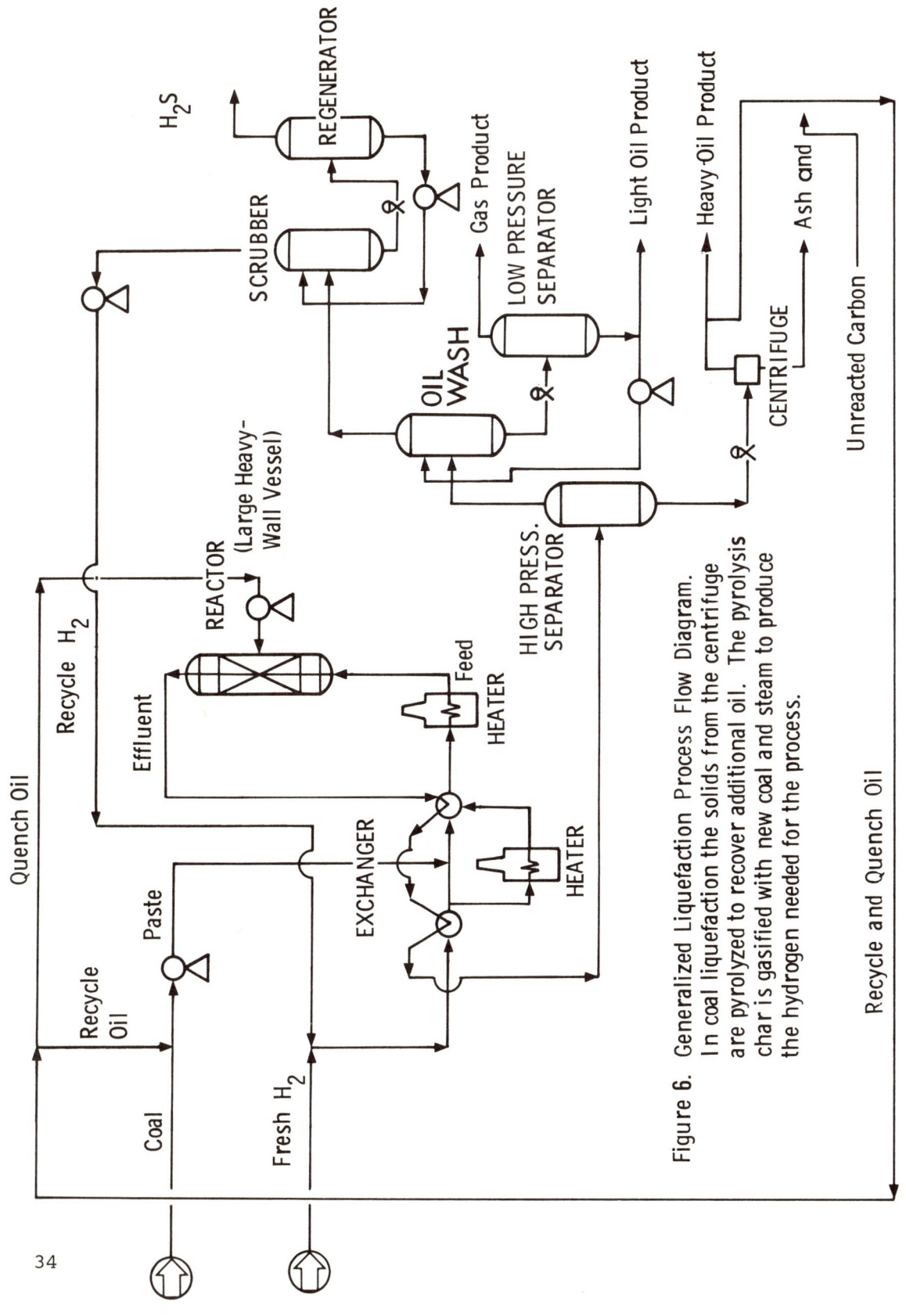

Figure 6. Generalized Liquefaction Process Flow Diagram. In coal liquefaction the solids from the centrifuge are pyrolyzed to recover additional oil. The pyrolysis char is gasified with new coal and steam to produce the hydrogen needed for the process.

operations and materials engineering. These are expected to be of an evolutionary character that should not require extraordinary efforts to achieve some scientific or engineering breakthrough.

RELATED ISSUES

While the construction of liquefaction plants will command engineering manpower and skilled craftsmen now in short supply, the major difficulty foreseen lies in the lack of fabrication capacity for certain highly specialized equipment such as reactor vessels, compressors and pumps. At present, there are about ten firms in the world capable of building the type of reactor vessels required--but about 50% of their capacity is now committed to the fabrication of nuclear reactors. Similarly, only one company in this country has had extensive experience in the fabrication of large high-pressure slurry pumps.

A full-scale coal conversion industry would exert a major impact on the supply of critical materials such as stainless and low-alloy steels, in forms such as heavy plate, forgings, castings and pipe. For example, between 15 and 20 thousand tons of steel would be required for each 50,000 barrel-per-day coal liquefaction plant. A major portion of this tonnage would be in low-alloy and stainless steels.

PRODUCTION CONSIDERATIONS

The overall thermal efficiency of the coal liquefaction process is about 85%. This is computed on the basis of the energy contained by the entering coal, the energy expended for operating the process, and the energy content of the liquified coal output. Although considerable energy is expended in separating the hydrogen from water, that same energy (less inefficiency) is retained in the output fuel.

The hydrogen required for plant operation should be produced by using the char byproduct of the coal liquefaction process itself for thermal input, plus additional coal and steam. Existing hydrogen-producing processes such as those of Texaco, Shell, Koppers-Totzek and Lurgi are available. Presently, worldwide, approximately 200 synthesis gas generators employing these processes are in successful operation using some type of petroleum stock as feed. The most suitable of these would require adaptation to coal and steam feed.

RELATED MATERIALS AREAS

In addition to the important erosion-corrosion study requirements reviewed above, a number of programs now underway to solve potential problems with materials for the hydrodesulfurization processes are directly applicable to coal liquefaction programs. More time than was available to this committee will be required to determine whether additional effort is necessary to meet the needs of the coal liquefaction program. The specific areas of concern include temper embrittlement of chromium-molybdenum reactor steels, non-destructive materials testing and inspection techniques, effects of hydrogen on chromium-molybdenum reactor steels, engineering parameters of two- and three-phase flow, and sour water corrosion. (See Appendix for details.)   MATERIALS PROBLEM

INFORMATION EXCHANGE

In addition to the areas of research listed above, adequate mechanisms must be established to expedite the acquisition, analysis and distribution of information needed by the coal liquefaction programs. Failures and their analyses should receive particular emphasis. Especially cited should be non-proprietary, unique construction details (weldments, joint designs, etc.); quality control or quality assurance programs; and the like. Encouragement should be given to appropriate professional societies to contribute to this effort. In some industries, this essential element of information exchange is effected through professional societies and industrial associations.

Oil Shale

Oil shale is expected to contribute 1.1 Quads in 1985. No serious material problems have been identified.

FORECASTS OF VOLUME

On the basis of known announced actions by industry and the Federal Government, oil shale production by 1985 will range between 500 and 750 thousand barrels a day. At this level, the basic constraint is economic viability as dictated by world petroleum prices. Technical risks also will have to be assumed by the developers. Beyond the 750 thousand to a million barrel-a-day level, additional constraints will become apparent, including for example air pollution standards in Colorado, water availability, future leasing practices of the Federal Government, and the problems accompanying a large, new industry in a sparsely inhabited region.

$\Delta Q = 1.1$

PARALLEL DEVELOPMENTS

There are various concurrent developments in oil shale technology. These are the Tosco-II process; the Union Oil Company SGR (Steam Gas Recirculation) process; several gas combustion processes such as those of the Bureau of Mines, the Petro 6, and Paraho processes; and *in situ* processing performed by several governmental and industrial entities. Each of these technologies is at various levels of development. The Tosco-II process is the most advanced towards commercialization; *in situ* is the least developed and requires more investigation. These technologies are being advanced in parallel by various industry groups as required for commercial development on their leased tracts. In fact, these parallel developments are required because more than a single technology in any commercial development is necessary to evaluate second-generation directions. The minimum project size that is now commercially and economically viable is a 50,000 barrel-per-day plant, made up of 6 unit trains each producing 10,000 to 11,000 barrels per day. Capital costs are estimated at $11,000 barrel-per-day capacity, equivalent to $5.2 billion per Quad.

While there is no existing "industry," several companies have announced plans, have performed engineering design, have generated environmental impact analyses, and are seeking pipeline and construction permits for their operations:

Colony Development operations on the Dow property for a 50,000-barrel-per-day plant scheduled to be operable in 1978; Union Oil Company on their private property for 50,000 barrels for their Steam Gas Recirculation process to be operable in 1980; the Federal CA tract with Gulf and Standard of Indiana who have announced operability in 1980-81 at 100,000 barrels per day, eventually rising to 200,000 to 300,000 barrels per day; the Colony group of Federal tract CB who have announced operability in 1981-82 at 100,000 to 150,000 barrels per day; Sun, Philips and Sohio on tracts UA and UB combined, to be operable in 1981-1982 at 100,000 barrels per day; and Occidental Petroleum has announced some probability of being operable in the 1978-80 period with 35,000 barrels per day from their underground *in situ* retorting method.

MATERIALS ASPECTS

There are no identified problems in critical materials or materials technology associated with oil shale development, nor any serious materials supply problems, other than long delivery lead times for large steel retorts and heavy equipment. As development proceeds, experience will undoubtedly identify materials problems, very likely of a corrosion nature.

The material most vital to the construction of oil shale plants is wide plate carbon steel for large retort vessels. The present shortage of steel is caused by increased demand, limited construction capacity, shortage of alloy materials and industry concentration on production of high-profit items. The most likely case of production of shale oil by 1985 requires a maximum of less than 1% of the 1973 steel product. However, fabricated structural steel now requires an average of 50 weeks to deliver--and this lead time is expected to increase by the end of 1974. Critical shortages of other required materials are not anticipated since these would be required in nominal amounts. Development time for oil shale processes could be substantially reduced by providing priority delivery of plate steel.   **MATERIALS PROBLEM**

EFFICIENCY

The general efficiency of oil shale processes for mining operations, crushing, retorting and slight upgrading is approximately 65-68% of the total energy processed through the plant. This efficiency takes into account all the purchase of energy and the oil shale which is consumed internally. For underground mining, approximately 65% of the energy in place is recovered, leaving 35% in the form of pillars which is not utilized. This latent energy is still available for potential recovery by later technology.

High Temperature Gas Turbines

INTRODUCTION

Gas turbines are in wide use today in aircraft power plants and in electric power generator stations. The gas turbine power generator has the lowest capital cost per kilowatt capacity of any fossil fuel system available now. High cost petroleum fuel is required for operation, so the major use of these units is in peak shaving that supplies the highest portion of the daily power cycle (see Figure 7). The efficiency of the gas turbine is directly proportional to the inlet gas temperature, so that high temperature materials development is the central issue in developing higher efficiency convertors. Programs now underway will just start to affect the energy situation by 1985, but should expand rapidly in the few years following. The two developing areas of importance are:

- gas turbine electric generators for intermediate duty, and

- automotive gas turbines.

INDUSTRIAL POWER GENERATING GAS TURBINES

> *On the basis of the projected installations of intermediate-duty electric generators, power systems using high-temperature turbines of greater efficiency would save about 0.05 Quad in 1985 (projections for 1990 indicate an annual national fuel savings of about 0.3 Quad). Achievement of this goal will require considerable materials technology.* $\Delta Q = 0.05$

In order to achieve the highest possible efficiency, the gas turbine will be used in a combined cycle where the hot exhaust gas from the turbine supplies heat to a steam turbine system which will also turn electric generators. At the same time, the turbines will be designed to operate on feed from gasified coal rather than on petroleum products. The use of gasified coal will compound the material development demands because (1) the particulates entrained in the coal gas even following separation) and (2) the impurity content of the coal gas will corrode blade material and cause loss of strength and rigidity. In the industrial power generation field, three distinct approaches are underway to achieve higher operating temperatures of the gas turbine--and thus improved efficiencies.   MATERIALS PROBLEM

Improvements Through Metallurgical Development

Historically, advances in superalloys and metal design and fabrication practices have led to continuous improvements in gas turbine efficiency by allowing gradual increases in operating temperatures--at a rate of about 10°F per year. It is expected that as a result of improvements in alloys, increases in operating temperatures will continue so that by 1985 an ultimate temperature of about 2200°F will be achieved. Such a temperature would allow about a 7% increase above the present efficiency (from 29% to 32%).

The investment in materials development (in advanced alloys, protective coatings and composites) to achieve the above goals is estimated at about $40 million, spread out over the next 8-10 years.

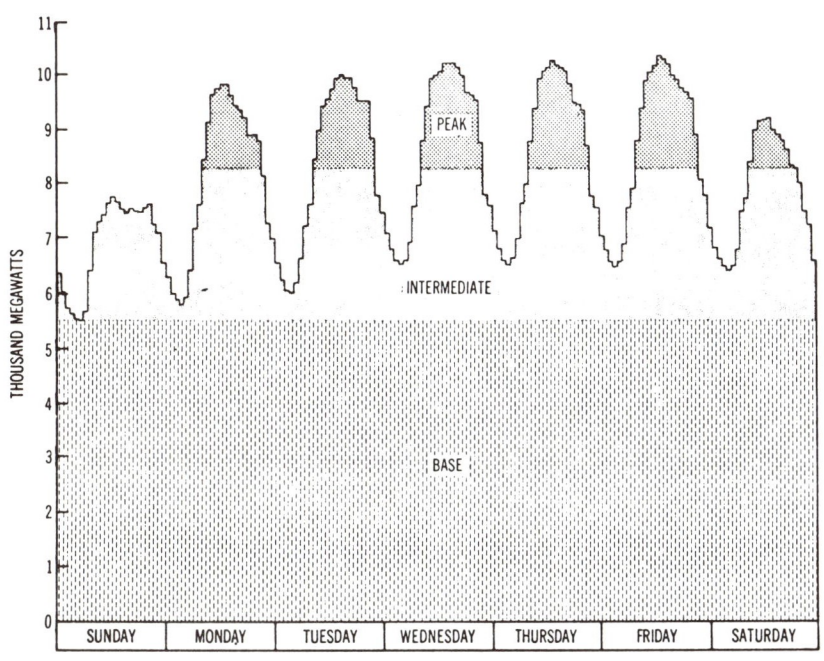

Figure 7. Typical load curves for electric power generation for one week. When additional generators are turned on to meet the peak load, the process is referred to as "peak shaving." Oil-fired turbines supply much of this demand.

Ceramic Gas Turbines

Engineering evaluations of new high temperature ceramics indicate that these materials, when used in the hot sections of gas turbines, will permit operational temperatures of 2500°F or higher. They will require no air cooling of critical parts and will consequently avoid the cooling requirements that penalize conventional gas turbines. At a temperature of 2500°F, the efficiency for combined cycle operation would be increased from 42% to 49% (a 16% increase over the 42% base). Improvements sought in ceramic materials characteristics include upgrading of strength in excess of 100,000 psi and of creep resistance at 2500°F. Forming and rapid machining processes for these ceramics will also require development. **MATERIALS PROBLEM**

It must be recognized that development of stationary and rotating turbine blades made of ceramic materials represents a complete new technology. We lack knowledge and experience in design criteria for some applications of brittle materials. Designing with ceramics is in no respect similar to designing with metals. The success that has been attained so far in ceramic turbines has been due to extremely careful and precise load and stress analyses using specially developed three-dimensional codes and high speed computers. New concepts for dealing with failure mechanisms based on probability analysis have had to replace the deterministic methods which have dominated designs using metals. New ways must yet be developed for the treatment of creep and fatigue in these materials over long periods at high temperatures and stresses. Few universities or industrial organizations have any expertise in this emerging material area. **MATERIALS PROBLEM**

The materials portion of the ceramic turbine development will require about $30 million over the next decade.

Fluid Cooled Gas Turbines

The development concept in this approach for improving the performance of gas turbines is based on the water cooling of critical hot section components rather than conventional air cooling (inlet gas temperatures of about 3000°F are expected). Some efficiency losses are incurred as a result of cooling procedures, but combined cycle evaluations indicate that achievable efficiencies would be similar to those for the uncooled ceramic system. Materials selection problems are not severe, but fabrication of the blade ducts required for circulation of the cooling fluid is complex. The development is estimated to be commensurate in time and cost with the ceramic gas turbine previously described. **MATERIALS PROBLEM**

THE AUTOMOTIVE CERAMIC GAS TURBINE

Fuel savings from this program in 1985 could be 0.08 Quad and 0.7 Quad in 1990. An aggressive ceramic engineering development is required. ΔQ=0.08

In 1985, the transportation sector will consume about 28% of all energy; yet only about one-quarter of that will be converted to useful work, the remainder being rejected as heat. Thus, the energy loss in transportation is at least as large as in electric generation, and the payoff in improved conversion is just as great.

Just as in power generating turbines, new high temperature ceramics are applicable in small vehicular engines. By operating at inlet temperatures on the order of 2500°F, improvements of about 40% over current internal combustion engine efficiencies are projected (i.e., an improvement from about 28% overall to 39%). On the basis of an average improvement of 9 miles per gallon for an annual 10,000 miles per year, and with the production capabilities projected by a potential manufacturer, the annual fuel savings in 1985 would be about 0.08 Quad and about 0.7 Quad in 1990.

A joint government-industry program has established feasibility through the stationary test stage. Emphasis is on designs which can be mass-produced at low cost. An inlet heat exchanger removes ingested objects. The remaining materials problems are substantial, however, requiring approximately $25 million to be expended during the period 1975 to 1981. An aggressive overall program could introduce about one-half million turbine automobiles in 1986, with increasing numbers in subsequent years.

Further details about automotive gas turbines are included in the Appendix.

Critical Elements

Critical elements do not in themselves directly contribute to energy production or conservation, but their availability is essential to other programs.

ENERGY PROGRAM REQUIREMENTS

The proposed energy program implies a pattern of substantial requirements for specialty materials. These requirements, superimposed on those routinely projected for American industry, are concentrated within the steel industry, and within that sector, on materials which can tolerate high stresses and temperatures, severely corrosive and erosive environments, and various combinations of these.   **MATERIALS PROBLEM**

Analysis of these requirements by the committee led to the identification of nine materials that warranted review as to adequacy of their supply to meet this expanded U.S. demand. Although precise quantitative assessments could not be made without exact, time-phased-requirements figures (and these are not available), it was possible in a general way to establish relative degrees of the need for advance action to improve supply-demand balance to meet future construction requirements of the energy program. Specifically, the materials were evaluated against such criteria as:

- essentiality for the energy program
- extent of reliance on imports
- potential adequacy of substitutes
- vulnerability of imports to concerted control of price or flow.

On this rough basis, an approximate ranking of the materials in order of their "criticalness" was determined as follows:

1. Manganese
2. Chromium
3. Fluorspar
4. Nickel
5. Cobalt
6. Aluminum (including bauxite)
7. Tungsten
8. Platinum (including platinum-group metals)
9. Copper

Emphasis was on the short-range future (1985), so consideration of projects to develop new alloy systems or substitutes involving extensive redesign of systems was excluded.

Main attention was focused on specialty requirements of the steel industry. One surprise was the appearance of fluorspar as a rather seriously "critical" material--attributable to its essentiality in both steel and aluminum production.

Because the energy program has been identified as a national goal, it was assumed that this program would have prior rights to available supplies of scarce materials. Accordingly, conservation actions would need to be addressed to non-energy uses of materials or to general improvements in supply-demand balance.

For each of the nine materials, two types of actions were considered: (1) supply-demand corrections, and (2) "quick pay-off" research and development. Beyond the obvious step of increasing supply, actions in the supply-demand area include recycling, eliminating non-essential uses, substituting more available elements for alloys, recovering materials from other industrial processes, and using new processes requiring less critical materials.

MATERIALS PROBLEM

The research and development effort with 1985 pay-off includes extending the performance of new alloys, writing handbooks for high performance alloys with low critical material content, and construction of demonstration plants capable of improving the critical materials balance. Research and development program costs, which are estimated for the next 10 years total $95-125 million.

Complete details regarding types of action required in the critical elements are provided in the Appendix.

MATERIALS ASSESSMENTS

IN OTHER ENERGY PROGRAMS

Nuclear

Figure 1 projects a 1985 contribution from nuclear energy of 13.2 Quads, an increase of 12 over 1974. Materials effort on the steam pressure vessels is required to assure public confidence and construction authorization.

INTRODUCTION

One of the most important single achievements which must be realized for Project Independence is that by 1985 about 30% of the total electric power generation will be by nuclear power reactors, about a factor of ten increase over 1974. About 95% of this installed capacity is expected to be the conventional light water-cooled reactors; the remaining 5% will be gas-cooled reactors.

$\Delta Q = 12.0$

Many materials problems peculiar to the reactor are of interest to the materials community. However, it is the feeling of this committee that there is a large, competent diverse community already intimately involved with development and design, and with the identification of important materials problems. Therefore, in the short time available consideration was limited to one problem area.

The record of current operation of nuclear power plants is still not satisfactory because the average downtime is about 35% compared with the anticipated 15-20%. This excessive downtime seriously affects the power costs as well as the projected quantity of energy expected from these plants. Analysis shows that downtime originates largely from problems in the steam system rather than in the nuclear system. Since the average layman does not understand these systems or problems, public concern has arisen over the safety of nuclear power plants. This concern has resulted with problems with the licensing of new plants.

THE STEAM SYSTEM

The reactor core is the source of heat for the power generator. In a pressurized water reactor, this heat is conducted by water pumped under pressure in a closed circuit through the reactor core and through a steam generator where the heat is transferred to water in a secondary system. Thus, there is no interchange of the water which is pumped through the reactor with that which is converted into steam for subsequent passage through the turbines. The steam which passes through the turbines does mechanical work which is converted into electrical energy in an electric generator. It is then condensed back into water at a condenser by coming into contact with a surface cooled by an external source of water. This condensed water then is pumped back into the steam generator. Thus, water used for the steam generation is also in a closed circuit, mixing with neither the water which goes through the reactor nor the water which condenses it. A fossil fuel electric power generating system is very similar, the nuclear reactor core and steam generator being replaced by a boiler heated by coal, oil, or gas.

The difficulties which have led to the excessive downtime of some nuclear power plants have been largely non-nuclear in the circuit of the steam system which passes through the turbines. Such difficulties also occur in large fossil fuel plants, but when they occur in nuclear plants their larger capacity and the greater concern for safety usually causes a longer downtime. Because of the similar origins of difficulties in steam systems of both fossil fuel and nuclear power plants, any research and development which leads to improved performance in steam systems will be applicable to nuclear plants. Problems which affect the structure of the steam system have both chemical and mechanical origins and it is important not only to correct the causes but also to develop techniques for continuous monitoring and on-line detection and for evaluation of potential problem areas.

Reduction in steam system difficulties has been achieved by chemical means after the discovery that the water should be kept in a narrow alkaline range to avoid stress corrosion within the system. One manufacturer, for example, achieves a condition of alkalinity with a mixture of disodium hydrogen phosphate and trisodium phosphate. With the ratio of sodium to phosphate ions kept within a certain limit this additive keeps calcium compounds in a flocculent state and buffers against alkali corrosion which can occur from sodium hydroxide buildup. Although a significant improvement, chemical additives have not completely eliminated corrosion effects because concentration outside of the optimum limits occurs as a result of local flow patterns of scale deposits inside the system.   **MATERIALS PROBLEM**

When such regions are predictable or occur with regularity, design changes can improve the situation--but problem identification can only come through accumulated experiences. An acceleration of the solution to this type of problem is possible with more laboratory research on the nature of stress corrosion, in particular within the chemical and physical environments actually experienced. It is realistic to expect that an improved understanding will lead more quickly to techniques of prevention.

A significant offender in the steam system is the condenser. This equipment, adapted from fossil fuel plants, has not received the detailed design consideration devoted to the nuclear components and its selection is often left to the contractor-builder rather than specified in the initial design. Yet defects here transfer contaminants from the outside cooling water to the steam system, leading to early trouble. While research on condensers may seem trivial, a vigorous effort to eliminate imperfections in this part of the system could significantly reduce the current excessive downtime.

The welding of the pressure vessel steel plates is done with care and with close adherence to critical standards. The plates, forging and welds are thoroughly tested over their entire surface. Although high standards exist, and pressure vessels are considered to be extremely safe, further knowledge is needed in welding and in inspection techniques. Improved reliable methods of testing for small imperfections after startup of a reactor and a technique for continuous inspection would significantly improve confidence in estimating safety and reliability. Such on-line monitoring is within the realm of feasibility.

PRESSURE VESSELS

The reactor pressure vessels are currently made by welding together steel forgings, plates, and rolled rings. These fabrication practices for vessels are conventional and the results have been satisfactory from a performance standpoint, but there are manufacturer capacity problems. Other processes might be more satisfactory, such as continuous build-up by electroslag deposition methods or by the construction of prestressed systems. However, it is not evident that producers of such vessels could achieve any gain in delivery date commensurate with the costs of research to develop such fabrication procedures.

The importance of reliable vessels delivered on time is imperative to Project Independence. Realization of the nuclear energy combination will require an increased production rate of reliable reactor pressure vessels.

Some estimate that unless new pressure vessel manufacturing capacity is added soon, as many as half of the pressure vessels being installed in the United States power plants will come from overseas sources by 1985. An element of uncertainty results from this use of foreign materials and techniques, although the information from our experiences with domestic steels has been conveyed to foreign producers and great care has been taken to insure compatibility of foreign standards with U.S. design practice. An example of such concern is in the long-time interaction of radiation-produced defects with impurities in the steels. There has been no systematic study of radiation effects in metals with varying concentrations of impurity atoms to identify possible combinational effects. There is a possibility that removal of such impurities in steels used in an environment of high temperature, pressure, and radiation of a nuclear reactor could lead to improved reliability. On the other hand, complete removal of all trace impurities may not be desirable because some may have an unexpected beneficial effect. The uncertainty introduced by foreign steels results from the different mix of scrap which is used according to the local market and circumstance. To eliminate this as a cause of concern will require a systematic study of the radiation damage to steels with different combinations and concentrations of impurities. With this information, future pressure vessel steels can be completely specified with respect to upper and lower tolerable limits of the critical trace elements, and acceptance testing can attend to the subtleties of the materials system.

MATERIALS PROBLEM

Theoretical fracture mechanics is the framework which allows interpretation of specific engineering tests of failure due to flaws or defects and allows extrapolation to a broad range of situations. It also provides a way for the powerful laboratory techniques used on small samples to be brought to bear on reliability calculations for massive, expensive structures such as pressure vessels. A substantial increase in this type of activity at universities, government laboratories and industrial firms would lead, in the time frame, to a better understanding of the reliability of pressure vessel systems in nuclear power plants, a wider agreement on estimates of that reliability, and a higher degree of confidence in design calculations of nuclear reactor power plants.

An effort involving 20 principal investigators for a period of 5 years (about $10 million costs) should yield a significant improvement in this important area.

Energy Conservation Through Materials Management

In the use sector, energy conservation is estimated at 20 Quads. Materials processing and utilization will influence much of this through a multitude of small effects.

INTRODUCTION

Of all short range approaches toward energy independence, conservation of energy offers the most promise. Quick returns already have been demonstrated by such simple measures as industrial plant surveys of energy economy opportunities, automobile speed reduction, and household thermostat adjustment. Major opportunities for further energy conservation are offered by techniques of materials management. Examples include the recycling of the approximately 3 million tons of aluminum annually discarded in municipal waste, reduction in size and weight of personal and mass transportation vehicles, extensive use of fibrous glass and other thermal insulating materials to reduce industrial and residential heat loss, recapture of energy in the form of heat from plastics and cellulosic materials in municipal waste, and many others.

$\Delta Q = 20.0$

The opportunities for energy conservation through materials management are clearly tremendous. Only two specific examples are examined here, however, because energy conservation measures based on current technology present problems of detail extending across the entire spectrum of materials processing and fabrication. Moreover, research and development of new processes and alternate materials to conserve energy must necessarily grow out of individual industry projects, and cannot easily be generalized. For purposes of this study, it is sufficient to call attention to the opportunity, to the need for quick response to identified conservation opportunities on the part of specific industries, to the need for sustained motivation and effort, and to the need for systematic engineering studies. New materials may require on the order of one or two decades to move from the laboratory to commercial use, and if they are to offer energy savings, appropriate action must be taken immediately.

The possibilities for conservation of energy through improved material processing and more careful process control have been considered extensively in several recent studies to determine their potential impact in the near term. In general, these studies indicate that the reduction in demand could be as much as 10 Quads in 1980 and perhaps 20 Quads in 1985.

U. S. ENERGY CONSUMPTION BY SECTOR

The largest energy consumers offer the greatest potential for saving. Use of energy in the United States in 1972 by major categories is shown in Table 3. Table 4 displays a rank order list of end-use consumption of energy in the United States in 1973. The major headings there which offer the greatest opportunity for energy savings through materials management are process steam, direct heat, electric drive, feed stocks, and electrolytic processes. This group consumed 30 Quads in 1973.

TABLE 3. Energy Consumption in the United States by Use Sector in 1972.

Use Sector		Quads	Percent of Total Consumption
1. Residential:			
	a. Heating	7.0	9.6
	b. Other	3.5	4.8
	TOTAL	10.5	14.4
2. Commercial:		6.2	8.5
3. Industrial:		22.7	31.0
4. Non-energy:*		3.7	5.1
5. Transportation:			
	a. Passenger cars	9.4	12.8
	b. Trucks & buses	4.0	5.5
	c. Aircraft	2.0	2.7
	d. Other	1.6	2.2
	TOTAL	17.0	23.3
6. Electricity Conversion**		13.0	17.8
TOTAL CONSUMPTION		73.1	100.0%

* Largely feed stocks
** Includes only energy losses in generation, transmission, and distribution

There is a serious lack of information on energy use in the industrial sector. With the present inadequate level of reporting, it will be impossible to monitor energy-saving programs in the industry. The poor quality of information is illustrated by the fact that 1968 is the most recent year for which detailed information is generally available for the industrial sector (see Table 5). This sector in 1968 accounted for 41.2% of the total energy consumption in the United States. (Electric utility consumption has been allocated to each end use.) Six SIC (Standard Industrial Classification) groups accounted for two thirds of that consumption. The primary metal industries are the largest single classification, but the SIC definitions make it difficult to identify the actual energy consumed in the total cycle of recovery of ore from the ground through manufacturing at the mill level.

TABLE 4. Energy Consumption in the United States by End-Use in 1973 (Estimated).

	ENERGY CONSUMPTION	
End Use	Quads	Percent of Use
Transportation	18.7	25
Space Heating	13.5	18
Process Steam	12.2	16
Direct Heat	8.2	11
Electric Drive	5.8	8
Lighting	4.2	5
Water Heating	3.0	4
Feed Stocks	2.8	4
Air Conditioning	2.3	3
Refrigeration	1.5	2
Cooking	0.7	1
Electrolytic Process	0.95	1
Other	1.8	2
TOTAL	75.6	100

NOTE: This breakdown of consumption (taken from "U.S. Energy Prospects," National Academy of Engineering) differs from that in Figure 1 of this report where space heating and cooling is identified for only the residential sector.

TABLE 5. Industrial Sector Energy
Consumption in Quads.

[In 1968, this sector accounted for 41% of the
total energy consumption in the United States.]

Industry Group	Coal	Natural Gas	Petroleum Products	Electricity	Total Energy
Primary metal industries*	2.838	0.863	0.306	1.291	5.298
Paper and allied products**	0.467	0.341	0.211	0.280	1.299
Stone, clay, glass and concrete products *	0.406	0.449	0.087	0.280	1.220
SUBTOTAL	3.711	1.653	0.604	1.851	7.817
Chemicals and allied products	0.666	1.219	1.426	1.626	4.937
Petroleum refining and related industries	-----	1.012	1.589	0.225	2.826
Food and kindred products	0.263	0.593	0.134	0.338	1.328
All other industries (includes wood and wood products)	0.976	4.781	0.721	1.572	8.050
TOTAL	5.616	9.258	4.474	5.612	24.96

* Generally excludes mining
** Generally excludes logging

The section which follows discusses extractive metallurgy of iron, copper and aluminum, the three large energy consumers. Because of its very large tonnage (131 million tons in 1968), steel is the dominant energy consumer, although copper and aluminum are both significant because of the high specific energy content.

Table 6 displays 16 industries or products which accounted for an estimated 50% of the total industrial consumption in 1968. Nearly 32% was consumed in what might be considered as materials industries. From that table, one can see that iron and steel, paper and paper board, petrochemical feed stock, aluminum, cement, ammonia, ferrous foundries, copper, glass and concrete are all candidates for energy conservation through improved processing.

The production of material itself is energy-intensive, requiring about 13% of the total national energy consumption. An appreciable part of that energy is now wasted in unsalvaged materials in dumps. Some of that energy is recoverable through recycling energy-intensive metals such as aluminum and copper, and by direct burning or conversion to fuel of combustible waste. Those possibilities are discussed in a succeeding section entitled, Fuel From Waste. Also, as previously mentioned, detail consideration where reduction in energy demand through materials management may be identified, is given in the immediate following section (pages 57-61) entitled, Metals Extraction and Processing.

Another way in which energy may be conserved is through the incorporation into design of an improved material which allows the same function to be achieved with less energy expenditure.

TABLE 6. Sixteen Industries (or Products) Which Accounted for an Estimated 50% of the Total Industrial Consumption in 1968.

Industry*	Share of Total Industrial Consumption (%)
Iron and Steel**	13.6
Petroleum Refining	11.3
Paper and Paper Board	5.2
Petrochemical Feed Stock	4.9
Aluminum	2.8
Cement	2.1
Ammonia	2.0
Ferrous Foundries	2.0
Carbon Black	0.9
Grain Mills	0.8
Copper	0.8
Glass	0.8
Concrete	0.7
Meat Products	0.7
Soda Ash	0.7
Sugar	0.7
SUBTOTAL	50.0
Other	50.0
TOTAL	100.0

* Generally excludes mining operations
** Includes blast furnaces, steel works and rolling and finishing mills. Excludes mining.

Metals Extraction and Processing

Material programs were not readily identified which would enable large energy savings in the near future.

INTRODUCTION

Primary metals production in the United States consumes about 8% of the energy production and, of this, 88% is used by the copper, aluminum, and steel industries combined (56% for steel, 40% for aluminum, and 4% for copper). Uses of this energy can be broadly categorized by:

- mining and beneficiation of ores,

- primary smelting,

- refining, and

- primary fabrication.

$\Delta Q = < 0.1$

Table 7 illustrates the breakdown of energy for these three metals in terms of BTU per pound of mill product.

Because these are fully developed industries with large capital investments, changes are costly and evolve quite slowly. Nevertheless, since they consume so much energy, opportunities for reducing consumption should be explored.

In response to increased energy costs, environmental considerations, and various other forces, technological changes are being introduced at an accelerating rate. In the past, innovative technologies were developed because of a multiplicity of other factors such as the decreasing grade or increasing complexity of ore, increasing costs of manpower, pollution factors, etc., with the net result of requiring more energy per unit of product. A few examples drawn from the primary processing industries for the three major metals will illustrate how additional technology may now reduce energy requirements.

TABLE 7. Energy Required to Produce
Three Primary Metals.

	Steel	Copper	Aluminum
Estimated % of Purchased Scrap Used	1	20	8
	BTU/LB OF MILL PRODUCTS		
Mining and Beneficiation	750	21,000	18,500
Primary Smelting and Refining	9,630	17,300	91,000
Primary Fabrication	8,330	13,100	15,400
Total Energy Per lb.	18,710	51,400	124,900
Tons Produced in U.S. in 1970	80×10^6	1.7×10^6	4.0×10^6
Energy Consumed in 1970 in Quads	3.2	0.18	0.99

MINING AND BENEFICIATION

Metal mining is confronted with concurrent problems of rising costs of energy and the necessity to cope with ever leaner ore bodies. Economies of scale are currently being utilized by the industry, but many current plans to increase efficiency are being hampered by the shortage of modern, efficient equipment. Thus, improved utilization of energy in the mining industry by 1985 depends in part on the availability of steel and heavy equipment for application to improved mining technology. For example, in the copper mining industry, the effect on productivity of the large truck to replace the rail car has been as great as any other recent innovation.

In the beneficiation area, comminution (crushing and grinding) is a huge consumer of energy and is an area which seems an appropriate target for significant savings. On the order of a billion tons of rock ore is crushed and ground every year within the United States, requiring about 0.1% of our energy production. Most of the applied energy is expended in the generation of heat and sound. Also, excessive energy consumption is caused by unnecessarily fine grinding. One approach to significant saving in this area is better control (through automation) of the size reduction and separation necessary to liberate the mineral values from the waste rock. It also appears that support for research in new equipment such as ball mill improvement could effect industry-wide savings.

On certain mineral commodities, grinding and beneficiation can be reduced or avoided by resorting to leaching processes. The Bayer process is one example, although its primary purpose is beneficiation of alumina to be charged to the electrolytic cell. In copper mining, the ore body can be shattered by explosives and leached either in place or in dumps by bacterial or chemical action. The copper values can be recovered from the solution by either replacement in solution by scrap iron or by electrolysis. In copper production, comminution and beneficiation comprise about 24% of the total energy production of copper ingots. Increased research and development in solution physical chemistry and hydrodynamics as applied to *in situ* rubble, and work on fracture mechanics for controlled creation of *in situ* rubble by explosives, could cut in half the energy now required to produce copper (51,400 BTU/lb--see Table 7). However, industry representatives point out that *in situ* experiments are very expensive and any laboratory finding must be field tested. Industry supports as much field testing as felt necessary to assure metal supply, but research, specifically directed toward long-range supply or for energy conservation would probably not be supported. The next decade will probably see application of the leaching technique to a few favorable locations but any large impact will depend on results obtained in the early tests.

MATERIALS PROBLEM

A marked departure from traditional mining techniques will be the recovery of manganese nodules from the deep ocean floor. The nodules are hydrated oxides of manganese and iron that contain valuable amounts of nickel, copper, cobalt, and molybdenum. Lack of operating experience means that cost estimates must be very tentative. However, preliminary projections indicate that this is a very expensive mining technique in terms of tons of ore recovered. Both the capital investment and the operating costs appear to be several times as large as that for underground copper mining, which, in turn, can only be justified by relatively rich deposits. Until some operating experience has been gained, it will not be possible to say which of the mineral resources can be economically produced. As surface reserves of critical elements are depleted, the nodules will become more attractive (see Appendix on Critical Elements). With present information, it is not possible to say utilization of this source of minerals would result in an energy saving.

PRIMARY SMELTING

Cheap natural gas has been plentiful over the past 30 years, and the nonferrous metals industry has come to rely on it as a fuel. The first inclination of the industry has been to substitute fuel oil for gas, but with petroleum shortages anticipated, a logical recourse is to return to coal as the primary source of energy for nonferrous metals smelting. The coal in this case may be converted to gas or (in reverberatory furnaces and other systems where ash and sulfur can be tolerated) burned directly in powdered form. Smelters handling sulfide ores and concentrates already possess, or will soon be required to possess, facilities for removal of sulfur oxides from stack gases, and hence the higher sulfur coals can be used with no penalty. Oxygen utilization can also play a significant role, and research in pyrometallurgy directed toward energy saving will be a natural part of the development of new fuel sources.

For aluminum, the primary smelting process is the Hall process, based on the electrolysis of alumina in a solution of a molten fluoride compound. This process consumes 72% of the 124,000 BTU required to produce one pound of aluminum ingot. The Hall process is not particuarly efficient, and much energy could possibly be saved through processes closer to theoretical efficiency. Fuel required to provide electricity for the Hall process contains seven times the theoretical energy required to decompose aluminum oxide. In the chloride process, which is currently being introduced as a successor, the alumina is chlorinated under pressure in the presence of carbon and converted to aluminum chloride. Fused salt electrolysis of this chloride results in aluminum metal, with a reduction energy consumption by 30% of that needed for the Hall process. Introduction of the chloride process, however, will be slow and will apply only to new plants. At the same time, significant energy savings could be made possible by increasing the efficiency of energy utilization in the Hall process still used in existing smelters. Examples of effective research and development areas include development of improved heat balances, bath chemistry and electrode materials. Estimates are made that energy savings of up to 15% may be possible by this route.

MATERIALS PROBLEM

Improvements beyond this primary smelting operation are also possible and may significantly reduce the fraction of energy (12%) used in remelting and fabrication of primary aluminum. The design of remelt furnaces, fuel optimization, and recovery from oxide skim and dross are all possible areas of improvement.

PROCESSING

The ferrous metal industry offers the greatest potential for energy savings by improved processing simply because it is by far the largest user. In the past 10-15 years, the iron and steel industry has already achieved significant reduction in energy requirements by replacing the open hearth furnaces by basic oxygen steel-making. A newer technology which is now being introduced in the industry is continuous casting. In the older plants, steel is heated an average of three times in its trip from the furnace to the final product, but in continuous casting, most of the first heat is retained with a saving of about 2% of the total energy required to make the steel. Because the process is applicable only to certain types of steel, and because small orders cannot be economically handled this way, not all steel-making capacity should be considered for conversion to continuous casting, but the present level of about 10% might be increased to perhaps as much as one half of the capacity.

Blast furnace gas is a low grade fuel (about 95 BTU/scf). Part of the gas is used for heating purposes in direct connection with the plant operation, but is has been estimated that as much as 0.4 Quads may be wasted each year in the United States. A program to recover this fuel for process heat or electric generation would yield significant results in the near term.

PROFESSIONAL TRAINING

The university centers in this country responsible for training professionals in the field of extractive metallurgy have, for many years, recieved minimal support. The number of departments offering courses has declined, and the number of students enrolled at both graduate and undergraduate levels has declined substantially. Although in the extractive industries it is possible to achieve technical competence without an extensive academic background, the committee believes that a program concerned with the near-term energy problem should also provide for university research in this field as well as the training of new mining and extractive metallurgical engineers.

Fuel From Waste

Energy recovered from the combustible fraction of waste is estimated to be 0.5 Quad, that saved by recovering metals 0.3 Quad. Materials effort will help assure success.

INTRODUCTION

The energy content of the total waste produced in the United States each year is about 12% of the total energy consumed. Because of its distribution and associated transportation costs, only part of the waste produced can be considered recoverable (see Table 8). The waste can be roughly categorized into three types:

$\Delta Q = 0.8$

- urban (including residential, municipal and industrial);

- agricultural (including animal and crop wastes); and

- forest (including sawdust, bark, etc.).

TABLE 8. Yearly Organic Waste Generation in the U.S.

	Total (10^6 tons)	Recoverable (10^6 tons)	Energy Content* (Quads)
Urban	300	160	1.9
Agricultural	590 (200 animal)	26	0.2
Forest	55	5	0.08
TOTAL	945	191	2.18

* A ΔQ of 0.8 represents the committee's estimate of a realistic target for 1985. The totals in Table 8 (2.18 Quads) suggest the ultimate possible recoverable energy.

The total urban waste is about 300 million tons per year, 160 million of which is potentially recoverable, representing an energy source of about 1.9 Quads. Since the total energy use of the country is about 70 Quads, this recoverable urban energy resource is about 3% of the total. Similarly, the total dried animal waste quantity is about 200 million tons of which 26 million is readily recoverable in large quantities from feed lots.

This animal waste contains a thermal energy resource of about 0.2 Quad. Crop waste has an equivalent recoverable energy, but is generally utilized agriculturally. Although 55 million tons of forest waste are produced in this country in one year, the use of this material for energy production amounts now to only about 0.08 Quad and is discussed only briefly at the end of this section.

ENERGY RECOVERY OF URBAN WASTE

Urban wastes are a mixture of combustible and non-combustible materials. While it is possible to burn unsorted urban waste, the value of the non-combustible components is then diminished, so it is preferable to have extensive separation processes prior to combustion. This waste may either be directly burned as a component in a fossil fuel furnace or pyrolized to a gas or fuel oil. Large scale pilot plants (about 20) using these options are now in operation or under construction.

Typical raw urban refuse contains metals, glass, dirt and combustibles in the approximate quantities in pounds per ton shown in Table 9. The total energy required to produce the metal and glass components from original sources (see Metals Extraction and Processing section) is 0.7 Quad which represents 1% of the annual total energy production of the United States. The energy for secondary refining, however, is only a small fraction of this. Approximately 60% of the metal and glass produced in this country does not reach the urban waste collection and is not reported in Table 9.

Economic operation of a refuse processing plant is enhanced if clean separation of the individual noncombustible components is achieved. Separation procedures of various types have been developed which are primarily based on long-standing technology of the mining and minerals beneficiation field. These processes invoke shredding, screening, air classification, magnetic and eddy current separation, gravity separation and flotation. The corrosive slags which cause the major corrosion problems in the incinerators are mainly due to the noncombustible materials which are kept out of the combustion chamber by these several processes so the life of the incinerator is extended.

MATERIALS PROBLEM

TABLE 9. Components of Municipal Refuse (Dry Basis).

[The energy content is a combination of that available by combustion and that which was originally expended to recover and refine metals.]

	Amount/Ton (pounds)	Annual Tonnage (millions)	BTU Content (Quads) or Consumption
Combustibles	1628	130.2	1.56*
Glass	128	10.2	0.02 ⎫
Ferrous Metals	154	12.3	.5 ⎪
Base Metals (Copper, Zinc, Lead)	6	0.5	.024 ⎬ .704
Aluminum	8	0.6	.16 ⎭
Ceramic, Dirt and Rock	34		
Glass Fines	42		
TOTAL			2.26

---

* Available for consumption

Source: U.S. Bureau of Mines

    The shredders used to comminute the municipal wastes are subject to considerable wear from the abrasive action of the refuse. Although the shredders are made with replaceable wear faces and can be reversed to present new faces to the refuse, maintenance of shredders is still very costly. The characteristics of the shredded product change as the shredder faces wear, causing difficulties in classifying, feeding and burning the material. Therefore, development of better abrasion-resistant materials for shredders would greatly aid both the recovery of the noncombustibles and the use of the combustibles for heat generation.

MATERIALS PROBLEM

Two other possible avenues of research are evident to solve the erosion problems in refuse feeders. One is to develop abrasion-resistant alloys for use in pipelines and feeder mechanisms--an approach of considerable appeal because such alloys would be of use in other areas. The other avenue would be to remove the glass more completely and thereby reduce the erosive nature of the feed.

A materials problem common to the burning of either separated or raw waste is the corrosion of metals and refractories caused by hydrochloric acid generated by pyrolysis of chlorine-containing wastes. Much of this chlorine originates in polyvinyl chloride (PVC) plastics, so the corrosion could be considerably reduced if an effective separation method for PVC could be developed. That corrosion rates of boiler tubes are partly a function of fouling is illustrated by the fact that after 1000 hours of operation of one trial boiler, 25% of the wall thickness had eroded on the final superheater tubes due to local coverage by PVC. Development of materials corrosion resistant to hydrochloric acid and iron oxides for use in refuse incinerators would accelerate the practical recovery of energy from waste. **MATERIALS PROBLEM**

MATERIALS PROBLEM

In the incineration of raw refuse, additional problems arise from the corrosive action of the slag which is generated. Incineration slags are very complex and contain nearly all the elements of the periodic table, the primary constituents being silica, alumina, iron oxides, and calcium oxides, with lesser amounts of titania and magnesia. Effects upon refractories are equally complex, partly because the slags are variable from point to point in the incinerators and from incinerator to incinerator. The largest portion of the iron oxide is present as ferrous oxide which is particularly destructive to refractories. This representative problem clearly indicates that effort is needed to develop refractories specifically designed for use in high temperature pyrolysis units. **MATERIALS PROBLEM**

ENERGY RECOVERY OF ANIMAL WASTE

There are three distinct ways of recovering energy from waste:

- direct burning,

- pyrolizing the waste to oil or gas by chemical change through heat, and

- enzymatic conversion to methane.

The preferred energy recovery process for animal waste is pyrolysis. Although the recovery of animal waste is still in its infancy, with only one pilot plant in operation, it has been learned that a conversion can be made of one ton of dry manure to net one barrel of oil. Thus, the potential recovery of 26 million barrels of oil annually represents about 0.5% of the 6 billion barrels of oil used annually.

This technology is applicable also to other forms of cellulose waste such as paper, wood and sewage sludge. The temperatures and pressures required are consistent with current conventional boiler technology and no materials problems are expected to arise. Another process under study involves low pressure pyrolysis which produces 12.5% char and varying quantities of oil, "tar", and gas depending upon pyrolysis temperature. Chemical products of the process include ketones and alcohols in the tar fraction and ammonium sulfate which is useful as fertilizer. Materials problems may arise here because of the combination of high temperature and ammonium sulfate.   MATERIALS
                                                            PROBLEM

An appropriate level of support in the materials area related to waste is $2 million per year.

ENERGY RECOVERY OF FOREST WASTE

Some geographical areas such as the Pacific Northwest, have large quantities of a wood waste called hog fuel. This fuel is frequently used in lumber mills to generate process steam. It may be used also by power companies to generate electricity for peaking loads. The technology for burning this fuel is well developed, and no new materials problems are apparent. Hog fuel is burned in traveling or fixed grate boilers. Some problem may exist in the traveling grate models because of dirt mixed with the wood or because of the somewhat abrasive nature of wood itself. Erosion may also occur in the pneumatic lines used to feed the waste to the boilers. Although the development of abrasion resistant alloys would benefit this technology, it is doubtful that there is sufficient need to warrant a large research effort.

Geothermal

Geothermal sources in California may produce 0.3 Quad by 1985. Considerable material development will be required to withstand hot water of high salinity.

INTRODUCTION

The heat conducted from the interior of the earth to the total surface over one year is equivalent to about 9800 Quads, but up to the present, only local "hot spots" have been exploitable to generate useful energy. Typical of these hot spots are geyser and hot spring areas where steam and hot water reservoirs predominate underground. These dry steam (steam vapor containing no free water particles) and hot water geothermal technologies present different materials problems, some as yet undefined.

$\Delta Q = 0.3$

DRY STEAM TECHNOLOGY

The United States has the world's largest known dry steam area in The Geysers field near San Francisco. Here, dry steam has been used commercially since 1960 to operate electrical generating equipment at costs very competitive with other power sources (see Table 10).

The potential for development of The Geysers field ranges from an estimated output by 1976 of 0.08 (0.9 GWe) to 0.13 Quad by 1985 (1.5 GWe) and up to 0.4 Quad as an ultimate. Another newly discovered field in New Mexico has yet to be assessed or commercially exploited.

A geothermal steam area is formed when hot magma, fairly close to the earth's cooled crust comes in contact with ground water that has percolated to depths of several miles and been converted to steam. If a layer of impervious rock or of deposited minerals seals off or caps the heat reservoir so formed, a natural steam boiler is created whose temperatures can vary between $400°$ and $700°F$, at a pressure up to 500 pounds per square inch. If the fissures rising from the reservoir are between 2000 and 9000 feet deep, these reservoirs can now be profitably tapped.

TABLE 10. Electric Power Costs for Geothermal, Nuclear, Hydropower and Coal Plants.

Electric Power Costs

	Geothermal	Nuclear	Hydropower	Coal
Plant Investment, $/kW.	$110.00	$225.00	$250.00	$150.00
Fixed Charges, 14%/year/kW.	15.40	31.50	35.00	21.00
Fixed Charges, mills/kWh.	1.95	4.00	6.10	4.36
Operating Costs, mills/kWh.	0.25	0.50	0.10	0.25
Energy Costs, mills/kWh.	2.66	2.00	----	3.00
TOTAL COSTS:				
Variable Load Factor, mills/kWh.	4.86	6.50	6.20	7.61
90% Load factor, mills/kWh.	4.86	6.50	4.55	5.92

NOTE: The plant investment figures given are based (except for hydropower) on the average for new tax-paying, privately financed plants ordered in the U.S. in 1970. The energy cost shown for geothermal is that paid by Pacific Gas and Electric Co. at The Geysers.

The steam supply contains trace elements of boron, sulfides and ammonia which could present serious problems, but the water is injected back into the reservoir, probably extending reservoir operating life as well as preventing chemical and thermal pollution downstream. Small amounts of other gases are liberated in the process but are vented in such small quantities to the atmosphere that they are not considered to be an environmental hazard requiring cleanup.

Given the well-developed technology (corrosion problems to plant equipment are dealt with by proper selection materials) and the relatively inert character of dry steam, the committee judged that no materials development would be necessary in the near term for dry steam geothermal exploitation.

SUPERHEATED WATER TECHNOLOGY

Liquid-dominated geothermal fields (in which the reservoir fluid is primarily water mixed with steam) are several times more plentiful than dry-steam dominated fields. (The Salton Sea area of California alone has a conservative potential of 5 Quads per year over 20 years.) Liquid-dominated fields present serious materials problems because of the combination of heat and minerals in solution. For example, the salinity content (depending upon the locale) may be up to 25% by weight (as in the Salton Sea brine fields) compared with 3.5% for seawater. The Salton Sea brine also contains chlorides of 13 other elements.

One way to circumvent the salinity problem is to "flash" the water; that is, to suddenly reduce the pressure so that part of the superheated water vaporizes immediately. The steam is available to drive a turbine; the minerals remain in the hot brine, but so does a great part of the heat (perhaps one-third of that usable). Far preferable to "flashing" is the direct use of the hot brine in the turbine, but this creates severe problems, primarily, those dominated by corrosion and deposition. Scale is deposited on iron surfaces of the well-head, steam pipes, separators, and turbines; corrosion attacks turbine nozzles, blades and shaft seals; turbine blades are eroded by blowout of fine rock dust from the well which exceeds the collecting power of the separator; on iron parts, coatings are formed composed of iron oxide, silica and magnesia. The main compositions of the deposited scale are soluble salts such as ferrous and potassium sulfates, hardened with silica.

MATERIALS PROBLEM

Corrosion can be reduced in some systems by neutralizing the effects of dissolved oxygen in the hot water with the addition of sodium sulfite. The only candidate materials corrosion-resistant to brines of the Salton Sea composition are certain ceramics, tantalum, some plastics and perhaps zirconium and Teflon, but these all represent materials application developments. Meanwhile, chrome 10% stainless steel, aluminum, and cured resins will remain the prime material candidates for evaluation and modification to meet the corrosion environment of the hot water systems.

The materials aspects of this evaluation and modification effort will require about $1.5 million per year for an extended period (5-10 years).

Solar Energy

Solar energy is projected in Figure 1 to contribute 0.1 Quad in 1985. Development of improved panel materials is required.

INTRODUCTION

Exploitation of solar energy is primarily a problem of adapting materials into designs of low-cost hardware to capture and store the diffused and intermittent energy from the sun. Sophisticated solar energy concepts have been proposed, but are unlikely to be realized within the next ten years. In the less sophisticated approaches, the problems are predominantly economic rather than technological. Only modest incremental improvements are envisioned in such straightforward applications as water heating and space heating in homes and offices. Since space and water heating are projected to use 15 Quads in 1985, these are the logical applications to receive attention. $\Delta Q=0.1$

Other possible approaches to "solar" energy, broadly defined, which were judged to have no significant impact in the early time frame are wind power, conversion of biomass into energy, thermal conversion to electricity, photovoltaic (direct) conversion, and systems to take advantage of ocean thermal gradients.

SOLAR ENERGY ADVANTAGES

Solar energy offers a number of theoretical attractions; it is free, clean, and available at the point of consumption. It is often most abundantly available at those times of day and year when electric power demand is at peak. Its use would directly replace oil, gas and electricity. Designs that capture solar energy for this purpose are usually modular, and can benefit from economies of scale in manufacture and installation.

DISADVANTAGES

On the other hand, solar energy presents a number of disadvantages. Being intermittent (annually, daily, and according to the weather), sunlight cannot be relied on as a constant source; it requires either a backup power source or else a combination of excess capacity plus storage. Because insolation is diffuse (on the order of 1 kw/meter^2 at maximum intensity), large arrays of interceptors are needed to accumulate useful quantities of sunlight--implying a very low-cost design in terms of dollars per unit of area. This requirement is amplified by the fact that sunlight comes to earth from continually changing directions so that capturing devices must track the sun or be large enough to be useful even at inefficient angles of incidence.

APPLICATION CONCEPTS

To incorporate any of several already existing technologies in new housing construction (at the current rate) would be technically feasible. One concept involves the use of solar collectors in the form of roof panels, fixed in place, which raise the temperature of a piped working fluid such as water. The heated water is then pumped to a system of absorption refrigeration for space cooling, into radiators or heat exchangers for space heating, or into a reservoir of hot water for household use. Several concepts of heat storage have been proposed, such as bulk heat storage (literally, a pile of rocks in the basement), a tank of eutectic salt, or some high thermal capacitance liquid.

CONSTRUCTION CONSIDERATIONS

To achieve an energy production of 0.1 Quad by 1985 will require substantial commitments in construction, as the following example illustrates:

> Assume that for each of the 6 years preceding 1985, 15% of new starts or 300,000 housing units (a mix of apartments and single family units) are built with solar energy provided 50% of the heating requirements of 100 MBTU/yr (that is, 50 MBTU/yr). The annual energy produced in 1985 by solar would be $(3 \times 10^5) \times 6 \times (50 \times 10^6) \cong 10^{14}$ BTU or 0.1 Quad.

At an average practical production of 0.3 MBTU/ft^2 per year for flat plate stationary collectors, each unit would require 166 ft^2.

Investment and operating costs are subject to considerable controversy. Costs depend on geographical location, fraction of load supplied by solar, and other factors. Present estimates for total system cost range from $10-20 per square foot of collector. Thus each unit would require about $1700 to $3300 investment to produce 50 MBTU/yr or 34-66 $B/Quad. Unit cost estimates for energy, based on maintenance, depreciations and interest cost of capital investment, are in the range of $4-10/MBTU.*

MATERIALS PROBLEMS

To achieve large-scale solar energy usage without government subsidy or mandate will require substantial manufacturing cost reduction. A large market will, of course, bring about an efficiency of scale. The collector itself is a major fraction of the cost, and from a materials point of view represents an opportunity for cost reduction by technology development.

To optimize collection efficiency, the absorber surface should absorb the solar spectrum but not emit in the infrared. Such a surface is called a selective absorber. For this, the most generally used materials for coating surfaces are copper sulfide and the Tabor black nickel (nickel-zinc sulfide). Both of these coatings are susceptible to corrosion in the presence of water vapor and may also be temperature-sensitive. Hence, better corrosion-resistant selective absorbers need to be developed. Economics could be improved by developing self-limiting electro-deposited selective coatings which give the desired optimum thickness. Development of thin (<1 micrometer) spray-on/paint-on coating systems could lead to significantly lower cost coatings. **MATERIALS PROBLEM**

Program costs for this type of material development will be approximately $20 million over the next decade.

---

*This may be compared with resistive electric heating at $3.6/MBTU if the electric rate is 12 mils/kWh.

Energy Storage

Energy saved in 1985 by electric power load leveling through batteries or flywheels will be negligible. Materials problems in the battery area are substantial.

INTRODUCTION

The high cost of oil and gas has stimulated interest in energy storage devices to replace gas turbines and other fossil fuel generating plants which are used in peaking and load leveling service for electric power generation. If the daytime peak loads could be supplied by electric energy generated over-night by the high-efficiency base load generators, overall fuel consumption would be reduced. Moreover, the fuel could be coal or uranium rather than fuel oil.

Pumped hydrostorage is now employed in limited service. Together with compressed air storage, hydrostorage promises to play an increasing role in load leveling in the 1975-1985 period. Neither requires any significant materials research and development effort. However, two energy storage devices involving considerable materials effort which will be of interest during the next 10 years are batteries and flywheels.

BATTERIES

Rechargeable storage batteries are anticipated to be the first materials-intensive energy storage system to be used for load leveling. The chief limitations of storage batteries for load leveling are high capital cost and short lifecycle and shelf life, compared with the usual 20-year expectation in the utility industry. However, it is also anticipated that with improvements in materials for the major functional components of storage batteries these difficulties can be overcome and batteries will significantly enter load leveling service by 1985. $\Delta Q = <0.1$

Another important application for batteries in the 1975-1985 time frame could be as the power source for vehicles for short-haul urban service. The chief limitations of batteries for electric vehicles are inadequate specific capacity and high cost. Development of advanced batteries with greater power and energy density, such as the sodium-sulfur or zinc-chlorine cells, are expected to also have major impact in this field at some future period. Technical objectives of the two primary battery applications, as compared with present technology, are summarized in Table 11. Oxidation-resistant electronic conductors are needed with adequate mechanical properties to support the positive electrode. New ceramics must be developed to seal the feedthrough. Better separators (a porous body which will absorb electrolyte) must be made to resist attack by both positive and negative active materials. Adequate performance must be demonstrated with inexpensive materials. Further details on this subject are included in the Appendix.   **MATERIALS PROBLEM**

TABLE 11. Battery Characteristics

	Technical Objectives		Present Technology		
	Load Leveling	Vehicular Applications	Lead-Acid Golf Car	Nickel-Cadmium	Nickel-Iron
Cost ($/kWh)	20	25	25	600	400
Efficiency (%)	75	50-70			
Life (years)	10	3-5	5	5	10-20
Life (cycles)	2500	300-500	400	300/2000	3000

FLYWHEELS

Flywheel energy storage presents materials development problems because materials for this purpose having very high strength-to-density ratios are demanded and are not now available. However, modern fiber composite materials developed for advanced aerospace applications do offer an ideal material of construction for flywheels. Design of flywheels also is important in maximizing the use of these relatively expensive composites. The application of flywheels for load leveling probably will not occur by 1985 because a considerable and necessary development cycle would not be completed by that time. However, the use of flywheels for vehicular propulsion, such as for trolley cars and buses, may see limited service by 1985, initially with steel flywheels and later with advanced composite flywheels.   **MATERIALS PROBLEM**

PROGRAM COSTS

Costs for materials research and development on the near-term battery systems will total about $70 million over the next decade. Flywheels will require materials support of about $15 million over that period to provide a demonstration stationary storage site utilizing advanced composites.

Fuel Cells

Energy or fuel saved by use of fuel cells in 1985 was judged negligible. An innovative materials advance could change this estimate.

INTRODUCTION

The principle of the fuel cell is reverse electrolysis in the presence of a catalyst by which a fuel and oxidant combine to generate direct current. It is another way to convert fuel to electricity and as such must compete with several highly developed alternatives. The fuel cell offers these advantages:

- silent, low-pollution, unattended operation;

- high efficiency at partial as well as full load; and

- economy in modular units to allow for distributed siting.

$\Delta Q = <0.1$

For applications where cost has not been a primary consideration (e.g., lunar vehicles), fuel cells have performed with considerable success and will continue to meet special needs.

On the other hand, present-technology fuel cells have some disadvantages: they must be supplied with fuel of high hydrogen content, and present systems are not economically competitive with other electric generators. However, the state of the technology is sufficiently advanced that a relatively small but admittedly speculative investment in a specific area of research could possibly remove the roadblocks so that fuel cells could have a large impact.

FUEL CELL SYSTEMS

Extensive studies have been performed and are underway on several fuel cell systems with potential for high power densities. These include the acid systems (e.g., phosphoric acid electrolyte); molten carbonate systems (e.g., lithium, sodium and potassium carbonate eutectic electrolytes); base (potassium hydroxide); and solid oxide electrolyte systems (e.g., stabilized zirconia and rare earth oxide systems).

Of these, major attention is being focused on the phosphoric acid electrolyte system and the molten carbonate systems that offer near-term practicality.

The Acid Cell

The phosphoric acid system requires a "reformer" or fuel conditioner to prepare a high-hydrogen-content gas from natural or process gas, naphthas, or distillates. Secondly, it requires a fuel cell stack in which the hydrogen is electrochemically reacted with air (at about 100^0C) by the catalytic action of platinum dispersed on carbon electrodes. Presently, the fuel cell stack produces direct current at about 200 volts which is fed to an "inverter" for conversion to alternating current.

An important part of the capital cost for the fuel cell is in the platinum catalyst. With present technology, the cost of the platinum catalyst is about $100 per kilowatt. The catalyst has an effective life of about one year at which time about 80% of the platinum is recoverable. Not only is the current cost of platinum high, but production is low and any increased demand could significantly drive the price upward. If fuel cells were to supply about 2.5% of the U.S. electrical generating capacity in 1985, this would tie up about 25 million grams of platinum; the total world production is about 50 to 70 million grams per year. With the many other anticipated increasing demands on platinum, e.g., automobile catalytic converters, petroleum catalysts, etc., substantial changes in the fuel cell platinum requirements will be required to make this generating system significantly cost competitive. **MATERIALS PROBLEM**

In addition, improvement in fabrication procedures must be developed for carbon electrodes of complex shape, utilizing binders and injection molding compounds which do not impair the effectiveness of the platinum. The materials problems associated with the reforming and electrical inversion components are minimal and not limiting. **MATERIALS PROBLEM** Scaling, however, from small demonstration models up to full production plants has not yet been accomplished, and no normal problems associated with such scaling should be anticipated. It is possible that a small penetration into the energy conversion market will be achieved by 1985.

The Molten Carbonate Cell

Because the electrode material itself has a catalytic action, the molten carbonate fuel cell does not require platinum. Molten carbonate fuel cell systems are similar to the acid fuel cell since input fuels must be conditioned and cell output requires inversion for alternating current applications. They differ in that the electrolyte must operate at eutectic temperatures which are relatively high (600°C) and electrodes of nickel or cobalt offer the most promise. The reactants for this cell are hydrogen and oxygen (air), producing water as the reaction product. They have the potential of utilizing a fairly wide fuel base (by conditioning) including oil, natural gas, and low BTU process gas, such as blast-furnace off-gas.

Single cells of these carbonate systems have performed for several thousands of hours with initially high efficiencies approaching 45%. Targets are similar long-term efficiencies of 45% (heat rate of 7500 BTU/kWh), but increased operating times on the order of 40,000 hours, with capital costs for installation on the order of $400 per kilowatt.

The major problem hindering the attainment of these goals, is, as in the acid fuel cell, the electrode life times. In many respects, the loss of effectiveness of nickel or nickel-base electrode catalysis is strikingly similar to the loss for platinum electrodes in the phosphoric acid fuel cell system. Attempts are currently underway to improve electrode performance by alloying and by forming composite structures. **MATERIALS PROBLEM**

The economic acceptance of molten carbonate fuel cells is directly associated with catalyst technology; the more metal used, the longer the lifetime, but the higher the capital cost. Experience has shown that catalyst lifetimes can be improved. In addition, the effects leading to catalyst degradation have been identified as the result of a coarsening of the dispersed metal which composes the catalyst. The reduction in effective contact area between the catalyst and electrolyte leads to electrolytic polarizations. These, in turn, lead to losses and poor cell performance.

Much metallurgical work on powdered and dense metals would seem applicable to electrode effects observed in fuel cells. This generic device for energy conversion appears to offer a unique opportunity whereby a modest research investment in dispersed metal electrochemical catalysts, with the goal of improving lifetime in service, could have a large pay-off. A five year program involving six principal investigators (not all at the same location) might solve the catalyst degradation problems or provide sufficient insight for early solution. It is mandatory that such a program include materials characterization expertise, utilizing all the modern examination tools (scanning electron microscopes, microprobes, X-ray, etc.), as well as expertise in the theoretical aspects of electrochemistry. The cost of such a program, in total, would be about $3 million. **MATERIALS PROBLEM**

Isotopic Separation of Uranium 235

The committee did not make an estimate of the energy saving in this area.

In 1973, the gaseous diffusion plants operated by the U.S. Atomic Energy Commission consumed about 1.5% of the electrical energy produced in the U.S. During that year, they operated at approximately one half of their design capacity. At the completion of the Cascade Improvement Program and Cascade Upgrading Program, which are now underway, the diffusion plants will use somewhat more energy or about 1.9% of the U.S. projected electrical generation in 1983. The energy used to enrich the uranium in gaseous diffusion plants is only about 3% of the energy produced by that fuel in power reactors.

Alternate processes have been studied intensively to provide separated uranium at less cost and particularly, in this context, less energy for operation. One promising process is the centrifuge. Estimates have been made that a given separative result through centrifuging can be achieved at 10% of the energy used in the gaseous diffusion method.

The ultimate limit of performance in the centrifuge separation process is the material strength of the rotating components. Avery and Davis, point out in their recent book that the peripheral speed of the centrifuge piece rotor is limited by the strength-to-density ratio of the material of construction. The most desirable materials they mention are glass fiber and carbon fiber. Other materials of interest are maraging steel and titanium.   **MATERIALS PROBLEM**

Another approach receiving considerable exploratory attention is laser isotope separation. The basic idea is to exploit the high resolution, high intensity capability of the laser to excite a U_{235} atom or compound to a state which can subsequently be readily acted on in a radically different way than the unaffected U_{238} fraction. Preliminary analysis indicates that the process could be highly selective so that perhaps only one stage of separation is required. Optimistic estimates suggest that the plant investment would be much less and the energy consumption would be less than 10% of an equal-capacity diffusion plant.

The technical details of these processes were not available to this committee. If technical feasibility should be proven in the next few years, the program could have an important impact in 1985. The materials problems cannot be identified until the process of choice is further defined.

OTHER ISSUES

This study has focused on the technological aspects of material problems related to the near-term energy program. The committee is confident that the materials community can resolve these problems if given the challenge and opportunity. Therefore, materials technology does not need to be regarded as the limiting factor in the energy program.

In those programs which may make major contributions to the supply of fuel, i.e., nuclear, coal liquefaction, coal gasification, and oil shale, the most serious near-term problem appears to be the delivery of materials. Even in the pilot plant and demonstration plant stage where time scales must be compressed in order to make early choices for production design, materials and components are requiring excessive lead times for delivery and even then do not always meet specifications.

Although the shortage of material and parts is wide-spread, by far the most important large scale needs in the energy program are for heavy steel plates, large forgings, and castings. Low alloy and stainless steels and nickel-base super alloys are the principal materials needed. Project Independence will create a major increase in the demand for these products. However, in 1974 the demand already exceeds the capacity for heavy plate by over a million tons per year. Assuming the normal historical growth in demand and a one million ton increase in capacity already announced by the plate manufacturers, the demand will still exceed supply by over a million tons in 1980 and 2-3 million tons by 1985.

If the committee's estimates on the demands for heavy steel are reasonably correct, the objectives of Project Independence have no chance of being achieved unless steel plate-making capacity is promptly doubled.

It must be emphasized, however, that the time available to this committee did not permit an in-depth analysis of the supply-demand situation and a quantitative inference must be withheld until an in-depth study can be made.

Although the committee did not have the time to investigate in any detail, it is the belief of the committee that a potential supply-demand deficiency also exists in the large forging industry. Large forgings are required in the manufacture of large turbines, generators, and pressure vessels. A U.S. Department of Commerce task force on Materials, Equipment and Construction in their report (first draft copy) prepared for Project Independence Blueprint, came to essentially the same conclusion.

It is also possible that a supply-demand problem is developing for large steel castings, but this is not believed to be serious. The nation seems to be reaching fabrication capacity for heavy steel components, especially pressure vessels, and the assessment is here repeated that by 1985, as much as one-half of the large pressure vessels used for nuclear power generation may have to be imported from foreign sources.

Adequate planning for Project Independence will require much more detailed information about projected use of steel than is now available. It is not sufficient to consider total tonnages but rather it is necessary to identify the particular type of steel required, e.g., plate, forging or castings. Plate requirements should state thickness, width, and length, and the forgings requirement should state size and complexity of geometry.

Likewise, the heavy plate and forging manufacturing capacity of steel facilities within the steel industry should be identified as to size, shape and alloy grade.

The American Iron and Steel Institute is an essential source of information and knowledge to participate in developing this information. Since the overall problem is sufficiently complex, it will also require an organization with experience in large-scale computer modeling to properly deal with the trade-offs involved. (The Bechtel Corporation, for example, has developed such a model for the general energy construction outlook.) A similar study dealing with the details of the steel supply situation may well show that constraints due to steel making capacity could be the controlling parameter in determining the optimum near-term energy solution.

The fundamental problems associated with steel heavy plate in the energy industry suggest that renewed effort should be made to discover alternative metallurgical approaches to large pressure vessels.

Another resource essential to the success of the energy program is the engineering skill required for design and development. Of particular interest to this committee are the scientists and engineers trained and experienced in materials research and development. Large numbers will be required in the energy program over the coming decades. Because of the shift in national priorities, it is to be expected that a considerable number of materials experts will transfer from the space program to energy-related activities. Since this shift is just now in process, it is difficult to tell how many of the needs can be satisfied from this source, but certainly not all.

Significant new entries in the materials field can come only from the universities. It is of great concern that the undergraduate enrollment in materials has been declining for several years. The recent National Academy of Engineering study, "U.S. Energy Prospects," in analyzing the near-term energy program has pointed out the large number of engineers of all types which will be needed in the next decade.

Whatever programs are designed to meet manpower needs, they must include materials science and engineering as an important element.

RECURRENT RESEARCH THEMES

In reviewing the various possible near-term energy programs and the related materials problems, a number of common research themes emerged. These are described briefly below.

The combination of erosion and corrosion, particularly at elevated temperature environments, is a serious problem in practically all of the programs discussed in this report. Relatively little fundamental work has been done on this subject. Concepts are not well developed, and certainly there is no common fund of engineering knowledge and practice. In addition to the specific application problems described earlier, the successful long-term resolution of the nation's energy problem will require considerable effort to develop needed basic understanding.

The push toward higher temperatures to achieve greater efficiencies can only be met, in some cases, by various ceramic materials. The increasing use of these materials for sophisticated engineering purposes will require development of a broad base of understanding and increased training capacity to produce new appropriately qualified professionals.

All of the trends in the energy industry are in the direction of demand of higher reliability of design and operation. Increased plant size, increased capital investments, higher pressures and temperatures, increasing public awareness, and closer supervision by government agencies, all contribute to those trends. Increased reliability will require improved understanding of failure mechanisms of all types, more sophisticated techniques for nondestructive testing, and improved techniques of quality assurance.

This study has reemphasized the continuing importance of steel technology. The ever-increasing demands placed on steel structures, coupled with the requirement for higher and higher reliability, give rise to the need for more detailed understanding of the solid state at a rigorous scientific level so that problems of stress corrosion cracking and crack propagation can be dealt with pragmatically at the engineering level. The program described briefly in the section on nuclear energy will, of course, be directly applicable to other programs such as coal liquefaction and gasification.

Catalysis is another general topic that pervades the energy field. Although it is mentioned somewhat in passing in this report, catalysis is of fundamental importance in the petroleum industry, in coal liquefaction, coal gasification, and fuel cells, and may indeed play a role in some types of recovery of fuel from wastes. Although a great deal of effort has been expended on this subject, development of new commercial catalysts still is based on empirical techniques, whereas a more complete scientific understanding would allow the design of catalysts to meet specific engineering requirements.

FUTURE STUDIES

It was only with considerable self-restraint that this committee was able to restrict considerations to the near-term materials problems of energy programs. That only modest leverage is available to the materials community in the 1975-1985 time scale is evidenced by the fact that the sum of the Δ (delta) Quads listed in this report, with the exception of undefined conservation programs, total 17.2 Quads in 1985.

It is in the longer time frame that materials efforts along with other research and development can yield enormous dividends. With more years, time is available to develop understanding of phenomena, to build a community of knowledge in important new areas, and to resolve in a thorough way the multitude of material problems which inevitably accompany any new major technology. For example, materials research started in the late 1940's and early 50's on material behavior in a radiation environment, has been an essential element to the nuclear power program which contributed 0.7 Quad in 1973, and is projected for 13.2 Quads in 1985.

Future study of the materials issues in the energy program could provide a more comprehensive coverage of all of the approaches which are potential contributors. It is essential to look further into the future in order to identify materials research programs which may have to be started now in order to provide the knowledge when it is needed. In controlled thermonuclear fusion by magnetic confinement, for example, the first wall problem alone surpasses the intensity and complexity of materials problems faced in the fission reactor program. Providing a basis of understanding for engineering design will require many years of intensive experimental and theoretical work, and much of this must be accomplished before even the prototype systems are designed and built. There is an intense desire for conversion of solar energy directly to electricity, but economic systems will require a marked reduction in cost of the solar cells. Some of that reduction will be achieved by new production processing, and some by innovative systems based on deeper understanding of the photovoltaic processes themselves.

The methodology developed in this study would be helpful in setting these and similar problems in perspective, although one must recognize that the uncertainties in projections increase as one looks further into the future. But, based on the committee's limited experience, such a study should be invaluable in highlighting the materials areas which must receive adequate attention and resources if desirable long-range solutions to the energy problem are to be achieved in this country.

In addition, with more time available, it would be desirable to identify the relation between the various programs and their environmental impact, and to more precisely establish the effect of given energy programs on the overall economic and material resources of the country.

APPENDICES

STUDY PARTICIPANTS

The committee benefited from an overall perspective of the energy program as described in addresses given during the two-week session by:

Dr. Russell Drew, Science and Technology Policy Office, National Science Foundation

Dr. Alvin M. Weinberg, Director of Energy Research and Development, Federal Energy Administration

Dr. Walter Hibbard, Energy Research and Development Office, Federal Energy Administration

In preliminary planning meetings, Dr. John C. Fisher, General Electric, Dr. Jack Schantz, Resources for the Future, and Hans Landsberg, Resources for the Future, made significant contributions.

The two-week workshop was organized in the following way:

Session on Coal Gasification

Albert M. Hall, Battelle Memorial Institute, *Chairman*
*Robert I. Jaffee, Electric Power Institute
Elburt F. Osborn, The Carnegie Institute of Washington

William P. Haynes, U.S. Bureau of Mines, U.S. Department of the Interior
Howard Leavenworth, U.S. Bureau of Mines
Robert Lewis, Synthane Pilot Plant, U.S. Bureau of Mines
Richard Lucas, Virginia Polytechnic Institute and State University
Niranjan M. Parikh, IIT Research Institute
Elio Passaglia, National Bureau of Standards
Harry Perry, National Economic Research Association
Michael Raring, Office of Coal Research, U.S. Department of the Interior
Carl Schulz, Consolidation Coal Company
†Paul Shewmon, National Science Foundation
Roger Staehle, Ohio State University
John B. Wachtman, Jr., National Bureau of Standards
Richard A. White, Bechtel Corporation

Session on Liquid Fuels

Riki Kobayashi, Rice University, *Chairman*
*Robert I. Jaffee, Electric Power Institute
†Robb Thomson, Federal Energy Office

Andy DeCora, Office of Research and Development, U.S. Department of the Interior
Henry Frankel, Office of Coal Research, U.S. Department of the Interior
James E. Gantt, Development (R&D), Union Oil Products Process Division
Charles Holland, Texas A&M University
John Hutchins, Cameron Engineers, Inc.
Ralph P. Gulley, Gulf Research and Development Company
James J. Lacey, U.S. Bureau of Mines
Henry G. McGrath, Procon Inc. (Union Oil Products)
Paul Yavorsky, U.S. Bureau of Mines
Paul Wellman, U.S. Bureau of Mines

Session on High-Temperature Turbines

Richard J. Charles, General Electric Research and Development Center, *Chairman*
†Arthur C. Damask, Queens College

Raymond J. Bratton, Westinghouse Research Laboratories, Research and Development Center
†Maurice J. Sinnott, University of Michigan
Francis L. Ver Snyder, Pratt and Whitney Aircraft

Session on Critical Elements

James H. Bechtold, Westinghouse Research Laboratories, *Chairman*
Albert M. Hall, Battelle Memorial Institute
†Franklin P. Huddle, Congressional Research Service, Library of Congress

Richard Pitler, Allegheny-Ludlum Steel Corporation
Meir Carasso, Bechtel Corporation
Sheldon Wimpfen, U.S. Bureau of Mines
Lou Santone, U.S. Department of Commerce

---

*Member, National Materials Advisory Board
†Member, Solid State Sciences Panel

Other Topics

    Arthur C. Damask, Queens College, *Chairman*
    Richard J. Charles, General Electric Research and Development Center
    Franklin P. Huddle, Congressional Research Service, Library of Congress

Nuclear

    Richard Begley, Astronuclear Laboratory, Westinghouse Corporation
    Spencer Bush, Lawrence Berkeley Laboratory
    James S. Kane, U.S. Atomic Energy Commission
    †Donald K. Stevens, U.S. Atomic Energy Commission

Fuel from Waste

    Carl Rampacek, U.S. Bureau of Mines

Solar

    Ronal Larson, Office of Technology Assessment, United States Congress

Batteries and Flywheels

    James Birk, Electric Power Research Institute
    Davis Douglas, Gould Corporation

Fuel Cells

    John R. Foley, Pratt and Whitney
    Robert E. Meredith, Oregon State University

Extractive Metallurgy

    Merton C. Flemings, Massachusetts Institute of Technology
    H. W. Lownie, Jr., Battelle Memorial Institute
    William Drescher, College of Mines, University of Arizona

National Academy of Sciences Staff

    Donald G. Groves
    John C. Barrett
    Farlan Speer
    Susan L. Munro
    Jay Wentworth
    Peggy Reinheimer

---

†Member, Solid State Sciences Panel

COAL GASIFICATION

PERSPECTIVE

The installed nuclear plant capacity and forecasts were taken from several references (see Reference section following). The coal gasification projection for 1985 is from Figure 1 in this report. The earlier years estimate are from the Office of Coal Research.

GWe are converted to BTU/yr assuming 10,000 BTU/kWe and continuous operation. Coal plants are assumed to produce gas at 1,000 BTU per standard cubic foot.

Capital cost for the nuclear portion of the power plant was derived to be $170/kWe. Capital cost for the gasification plant is taken as $390 million for a 250 million standard cubic foot (SCF) per day plant.

BACKGROUND

Successful conversion of coal to clean, gaseous fuels depends upon efficient processing technology. Several commercially mature processes already exist for the production of low and medium BTU gas from selected coals. These must be improved and extended to accept a wider range of domestic coal types. Fuether, the efficiency of these processes must be improved to attract broad commercial acceptance by the U.S. electric utility and basic raw material industries. Successful commercial production of high BTU gas from coal has not been demonstrated. Even more than in the case of low BTU gas, these high BTU gas processes depend upon high temperatures and pressures to achieve reasonable efficiency levels.

These coal gasification processes impose severe service demands upon major components. Pressure vessels, pumps, valves, heat exchangers and piping and power conversion components at the combustion interface are especially critical. In addition to the severe mechanical and thermal demands of the high-pressure, high-temperature conversion processes, there are additional critical chemical factors. These arise in the chemical refining of the coal. Sulfur, alkali, metals, steam, chlorides, and other impurities must be dealt with under a broad range of operating conditions.

Problems are further compounded in most of the processes by hot erosion of the interiors of primary components. Fly ash, char, molten slag, and other solids must be transferred efficiently or entrained in rapidly flowing gases or liquids, or handled as hot massive solids.

As is well known, the gasification of coal is an old technology, and was practiced for many years as a cyclic process reacting coal in a fixed bed alternately with air and steam at atmospheric pressure to produce water gas, a medium BTU gas. Three coal gasification processes typify commercial use: Lurgi, a pressurized fixed bed gasifier; Koppers-Totzek, an atmospheric entrained bed gasifier; and Winkler, an atmospheric fluidized bed gasifier. These gasifiers share some features with advanced gasifiers now under development: operation under pressure, elevated temperature, and moving-bed reactors. It was anticipated that materials of construction could be selected on the basis of service experience provided by these established processes and by other industries, notably the petroleum industry.

Despite the above, the initial operation of the two small high BTU process demonstration plants, Hygas and CO_2 Acceptor, have been so plagued with operational difficulties and frequent shutdowns that the longest continuous run in either has been nine days. Many of the problems were the result of materials-design failures. Problems are anticipated in all pilot plants; indeed this is one reason why small plants are built before major capital is expended for a full size demonstration plant but the frequency of these failures forecasts the work yet to be done.

The materials problems observed in operating the Hygas and CO_2 Acceptor plants can be exemplified as follows:

1. It is difficult to introduce fine solids into pressurized reactors. The Lurgi process introduces sized lumps of non-coking coal into the reaction chamber through lock hoppers. In the Hygas process, the coal is pulverized, mixed with a liquid to form a slurry which is then introduced through valves. These wear excessively, and when the slurries are transferred at elevated temperatures, the wear problem is compounded by corrosion.

2. For injection or let-down, where dry solids are transferred hot, valves undergo excessive wear.

3. Transfer of fine solids at 50 feet per second by pneumatic means at elevated temperature results in excessive wear of liners, particularly at elbows and changes of direction. This has been a particular

problem in the CO_2 Acceptor process, where hot char and dolomite must be transferred at pressure between the regenerator and the gasifier. In fact, it will be true of most fluid bed processes. The cyclone separators also are subject to wear from high temperature erosion.

4. The ceramic liners of cast alumina refractory will deteriorate by steam leaching of the silica-base binding phase. The ceramic liner may crack and spall off as a result of plug formation in the fluidized bed. Ceramic linings have also been eroded by fine particles blown into the gasifier, requiring design changes.

The major materials problems that have been reported are erosion, corrosion (chiefly in valves and transfer lines) and spalling or leaching of the ceramic liners that have occurred in early stages of the operation. Other problems are anticipated such as carburization and sulfidation of metallic parts in cyclones and entry ports.

Although this listing of known problems is reassuring in some ways, one need only to mention some of the following unpredicated problems encountered in other major engineering endeavors to recognize the essential role of closely related materials science in the coal gasification program:

- the neutron-induced void growth in stainless steel exposed to liquid metal fast breeder reactor neutron energy spectra.
- the stress corrosion cracking of stainless steels in nuclear power plants.
- the hot corrosion in high temperature gas turbines.

COAL GASIFICATION PROCESSES

Coal gasification processes are commonly classified by the energy content of the gas produced--that is, low, medium or high BTU gas.

1. <u>Low BTU Gas (100-200 BTU/SCF)</u>

Low BUT gas is produced by the partial combustion of coal with air and steam. The gas consists of carbon monoxide, methane, hydrogen, and nitrogen along with carbon dioxide, water, hydrogen sulfide and other impurities, depending upon the particular process. Low BTU gas is intended for use at the site of generation and in today's practice is intended mainly for use in electrical power generation and process heat.

The operating characteristics of several experimental low BUT gasification plants are listed in Table 12. The various experimental gasifiers have fixed beds, entrained beds or fluidized beds. Some are pressurized and some are not. The low BTU gasifiers all have the high temperature, erosion/corrosion and structural instability problems of the high BTU gasifiers. A problem unique to the low BTU gasifiers is the necessity for hot clean-up (of sulfur and particulates) if this gas is to be efficiently used as a fuel for electrical power generation. Present scrubber materials cannot survive at the outlet temperature of the reactor, yet to cool the gas removes such an important fraction of the BTU content that overall efficiency becomes inadequate.

2. <u>Medium BTU Gas (200-500 BTU/SCF)</u>

Medium BTU gas differs from low BTU gas primarily in that it does not contain significant quantities of nitrogen. It is produced essentially in the same way as low BTU gas, except that oxygen-steam rather than air-steam is used in the coal gasification step, or air is used in a secondary combustor or regenerator and steam only is used in the gasifier, as in the CO_2 Acceptor of Union Carbide-Battelle processes. Raw medium BTU gas consists primarily of carbon monoxide, methane, hydrogen, with varying quantities of hydrogen sulfide, and other impurities.

In today's technology, medium BTU gas is of interest primarily as a raw gas for hydrogenation to methane or for liquefaction to synthetic crude. However, the gas may also be used on site for industrial processes or electric power generation. The thermal efficiency from coal-to-gas is not generally as high as for low BTU gasification. The energy content is not high enough for economic transportation over long distances by pipe line.

If intended as a raw gas for further hydrogenation to pipeline tas or liquid fuel, it is desirable that the gasification be at as high a pressure as possible so as to optimize the methane content in the raw gas. The generally higher pressures relative to low BTU gasification create unique requirements of strength, erosion and corrosion resistance, and resistance to hydrogen damage in the structural materials for this system.

The operating characteristics of several commercial and experimental medium BTU gasifiers are listed in Table 12.

3. <u>High BTU Gas (1000 BTU/SCF)</u>

The term "high BTU gas" usually refers to essentially pure methane intended to replace natural gas and to be distributed to industry, commerce, and residential consumers, through conventional pipe line systems.

In most gasifier schemes, the starting material is medium BTU gas as described above. However, some experimental gasifiers are built to directly hydrogenate the coal with added hydrogen, as in the HYGAS Process (see Table 12). Regardless, all processes require one or more catalytic steps. To prevent "poisoning" of the catalyst, the medium BTU gas must be cleaned of sulfur and particulates to a very high degree. Since the gas temperature must be reduced for the hydrogenation step, cleaning by wet scrubbing at a low temperature is acceptable but the possibility of aqueous stress corrosion is introduced at this point.

The Brookhaven National Laboratory estimates that as much as 150 million tons per year of coal may be converted to high BTU gas and 50 million tons per year to low BTU gas by 1985, at an estimated cost of $16-22 billion in capital expenditures. A typical plant might treat over 5 million tons of coal a year, cost $500-700 million and produce 250 million cubic feet of gas per day.

The materials problems in coal gasification systems are magnified inevitably by the tremendous size of some of the components in the proposed production plants. The gasification vessel best illustrates the size factor. A. M. MacNab of C. F. Braun and Company estimates that the size of these vessels can be as large as 22 feet in diameter and 250 feet tall, and they can weigh as much as 2,000-4,000 tons--as much as a large submarine--in the largest of these units. Even the "small" vessels in the high-pressure, high-throughput plants may be 12 feet in diameter, 170 feet tall, and weigh 500 tons--larger than the reactor pressure vessels in a nuclear power plant. An individual 250 million cubic foot-per-day plant may require as much as 24,000 tons of steel. It is estimated that on the order of 30 such plants will be needed by 1985; thus, some 720,000 tons of steel will be required, most of which will be in the form of heavy plate. Another 2,400 tons of nickel base alloys and stainless steels will be required per plant, or 72,000 tons for 30 plants.

The larger vessels will have to be field erected and will impose severe demands on the procedures by which they are welded and inspected. Improved welding techniques and inspection procedures may prove to be as important an area of effort as will be the effort to better define the performance characteristics of materials in the gasifier environment.

TABLE 12. <u>Selected Coal Gasification Processes</u>

Process	Reactor Bed Type	Gasifying Medium	Atmospheric Pressure	Temperature (°F)	Percent Efficiency	$Cost 10^6 BTU
LOW BTU GAS						
Commercial						
Winkler	Entrained	air-steam	1	1500	70/80	$0.9/ 1.25
Experimental						
Lurgi	Fixed	air-steam	20	1000		
GE Fixed Bed	Fixed	air-steam	8	1000		
Westinghouse	Fluidized	air-steam	10-16	1300-2000		
MEDIUM AND HIGH BTU						
Commercial						
Lurgi	Fixed	oxygen-steam	30-35	500-2000	56/68	$1/1.50
Koppers-Totzek	Entrained	oxygen-steam	1	1750-2350		
Winkler	Fluidized	oxygen-steam	1	1500-1800		
Experimental						
Hygas (IGT)	Fluidized	Hydrogen	75-100	1200-1800		
CO_2 Acceptor	Fluidized	Air-regenerator Steam gasifier	10-20	1575		
Synthane	Fluidized	oxygen-steam	40-70	1100-1850		
Bigas	Entrained	oxygen-steam	70	1st -2700 2nd -1700		
CO Gas	Fluidized	steam	1-3	1600-1700		
UC-Battelle	Fluidized	steam	6	1600-1800		

PROGRAMS REQUIRING ADDED EMPHASIS

Although considerable effort is already being expended on the coal gasification program, the committee is of the opinion that several areas will require increased emphasis if the production goals for 1985 are not met.

1. Materials Test Facility

Although the ultimate test for a material comes only in actual service, the problem for the next decade is the selection and evaluation of materials in such a way that service life will be satisfactory in full-scale facilities. Pilot plants will, of course, provide important materials test and evaluation data and it is tempting to assume that pilot plant experiments will be sufficient. An inevitable conflict arises, however, between the need for process information and operating experience on the one hand and materials evaluation on the other. The Atomic Energy Commission has demonstrated the importance of operation of a separate material test facility where material evaluation is given first priority. For the coal gasification program, such a test facility will necessarily be large and expensive because it should contain all the elements of a gasification plant; that is: coal feeding, gasification, clean-up, and catalysis; and furthermore, it must be of large scale because some of the materials problems result from plant size. The coal gasification material test facility will be a development challenge in itself because it must be capable of providing the most extreme operating conditions which may be faced by materials under actual operating conditions.

An important secondary benefit of the test facility will be the expertise which inevitably will be gained by the staff serving the facility. In addition, this will be a natural place for interchange of materials design information between the different individuals and organizations contributing to the coal gasification program.

2. Failure Analysis

Programs requiring exceptional performance with high reliability, such as the nuclear weapons program, the communication industry, or the space program, have developed a philosophy of complete and detailed failure analysis whenever a component fails in development or in production. Such analyses can be very expensive in money and professional time but are known to be invaluable in meeting difficult technical program objectives.

No such philosophy has as yet pervaded the coal gasification program. In addition, channels of communication musb be developed to transmit the results of such failure analyses to other organizations which may thereby avoid later failures of a similar nature.

3. Joint Materials-Design at All Levels

In the coal gasification program, primary emphasis to date has been focused on the processes which may prove satisfactory for large scale production. If full scale facilities with satisfactory reliability and operating life are to be in place in 1985, it is high time that the appropriate materials people be intimately involved at every step in the design procedure.

4. Standard Test Procedures

Coal gasification plants introduce a combination of environments which have not previously been experienced. Therefore, we have no background in developing standardized test methods, specifications, or quality control procedures. Examples of particular need are:

a. Erosion of metals and ceramics by ash in a stream of hot, corrosive gase,

b. Wear of metals and ceramics by coal and ash particles,

c. Hot corrosion of metals and ceramics in combustor and gasifier atmospheres,

d. Fused-salt corrosion of metals and ceramics,

e. Molten slag corrosion of metals and ceramics,

f. Thermal stress failure resistance of ceramics, and

g. Thermal conductivity of ceramics under gasifier conditions.

This is no simple matter, since the development of each test procedure will require sufficient basic understanding confirmed by full scale tests to correlate full scale long operating life properties with small scale test or evaluation.

5. Information Interchange

As has already been pointed out in the section on coal liquefaction, more effective channels must be established for timely, complete exchange of information between individuals and organizations.

JOINT MATERIALS/DESIGN DEVELOPMENT OF CRITICAL COMPONENTS

There are eight components in the coal gasification system which are critical to the operation and for which no satisfactory materials exist in all cases. The reliability of these components depends not only upon the development of satisfactory materials but also upon the design of the components so that the materials are not pressed beyond their capacities to perform over long time periods. These components and their special considerations are now described.

1. Gasifier Pressure Vessels

Pressure vessel design and fabrication is an old and complex art basic to many areas of concern in energy generation and conversion, e.g., nuclear energy, petroleum refining, steam boiler and electrical power generation. The ASME boiler and pressure vessel codes are the primary guides to assure the reliable and safe construction of these pressure vessels.* Although a very extensive library of knowledge exists for the materials selection and design principles for pressure vessels, the pressure vessels for coal gasification will be larger in size and will be exposed to more hostile environments than heretofore encountered so that our library of knowledge of materials property data will have to be significantly extended. It will have to be expanded to include new, stronger steels and other classes of materials such as reinforced concrete. Better materials properties data are needed on fracture toughness, creep, fatigue and other basic materials properties and especially on the magnitude of the deterioration of these properties in hostile environments.

In view of the very large size and critical functions of this component, special and very careful attention should be given to all aspects of the pressure vessels used for coal gasification. The vessel "system" considered should include the refractories, internal structural arrangements, the water jacket (if used), and penetrations. Factors considered should include:

a. Metallurgy (fracture toughness, temper embrittlement, weld overlays, and fatigue);

b. Ceramic materials (including the use of multi-layer composites, fastening to the vessel wall, prevention of deteriorationtion during operation, optimization of erosion resistance of thermal insulators, and resistance to damage from thermal stresses);

c. Non-destructive inspection (during both construction and operation);

d. Fabrication processes (involving plate manufacture, field assembly, welding, heat treating, and placement of insulation materials); and

e. Overall design criteria (including seismic considerations, maximum transient temperature and pressure, transient changes in gas composition).

2. High Temperature Particle Separators

Coal-derived gaseous fuels are intrinsically "dirtier" than conventional petroleum-derived fuels. Both chemical impurities and entrained solids (especially ash) are destructive to down-stream components, particularly those exposed to the combustion interface, e.g., boiler tubes, heat exchangers, and open cycle gas turbine components. Cooling of these "dirty" fuels to facilitate "clean-up" imposes severe economic penalties by loss of thermal efficiency. This makes hot clean-up a necessity. Materials involved in such hot clean-up components are highly vulnerable to the destructive effects of the impurities. For example, cyclones for centrifuging solid fly ash erode rapidly under these high-temperature, high-stress conditions. Filters and chemical clean-up processes are exposed to similar difficult problems.

Turbines used in combined cycle systems will not have economic lifetimes unless the particulate materials are removed in prior steps. Such methods of particle removal as cyclones, electrostatic precipitators, and filters must be developed for the temperature range of about $1800°$ F. By their nature, these erosion problems are expected to be severe and are aggravated by the combined effect of high temperatures, aggressive environments and high velocities. By reducing the concern for erosion in the turbine, engineers transfer the erosion problem to the particle separators. Specific considerations in the design of these units will be:

a. Minimal pressure drop across the unit,

b. Maximum effectiveness for the longest time,

c. Prevention of "break-throughs" where the particulates suddenly go directly to the turbine, and

d. Ready removal of the filtered particles across the pressure boundary.

Additional specific problems with the separator are the creep strength of cyclones, the charge transfer in electrostatic precipitators, and the clogging of filters. Particle separators are also required ahead of the methanator but the temperatures are lower (about $800°$ F). This technology is presently available from the steel industry.

---

*In view of the AEC's experiences in getting the Code extended, it is imperative to begin the process of getting the Code extended to coal gasification pressure vessels as soon as possible.

3. *Valves, Seals, and Controlled-Rate Feeders*

These components are required both in the low temperature portions of the system where there are large mass flows of solid materials and in high temperature portions where there are high velocity particulates entrained in gases. In addition, in both portions of the system, these components may be required to operate at extremely high pressures.

The essential problem in the use of valves, seals, and controlled-rate feeders involves the reactive particulate materials which cause wear and possible mal-operation of closely dimensioned working surfaces. From the design point of view, the objective is to keep these particulates away from working surfaces; from a materials point of view, the presence of some particles must be accepted and facing materials systems must be developed which resist the combined chemical and abrasive action.

While valves do not represent as large a capital investment as the pressure vessel, their mal-operation can shut the plant down as surely as a mal-operation in the pressure vessel.

There is a great incentive to develop a system for continuously feeding coal into the gasification system against a back pressure of up to 1500 psi. This process is presently handled by batch feeding of lock hoppers where alternate valving is utilized. The high frequency of operation of these sequential alternate valvings suggests that their lifetime will be severely limited.

4. *Catalytic Units*

High BTU systems depend upon the use of catalytic processes. These make it possible to upgrade the 300-400 BTU-SCF gas from the gasifier to the 900-1000 BTU/SCF value desired for pipeline quality. Possibly limiting problems which must be resolved for these units to operate reliably and economically are the following:

a. Deterioration of the catalyst which results from both the poisoning effects of impurities and the change in surface structure,

b. The pressure drop which results from the necessity to react the gas with the catalyst surfaces,

c. Refurbishing the catalysts (technique and rate), and

d. Mitigation of effects produced by exothermic catalysis reaction.

5. *Transfer Lines*

The large solid and gaseous mass which flows among the reaction vessels is carried in transfer lines. Essential problems which must be resolved here are:

a. Development of joints to allow for thermal expansion of the metal and ceramic systems,

b. Prevention of excessive wear at discontinuities (which produce flow eddies) and at bends,

c. Methods for non-destructive inspection of integrity, and

d. Procedures for repair of ceramic lining.

6. *Sulfur Removal at High Temperatures*

Where the produce of the gasifier is expanded through a turbine, the sulfur must be removed from hot gas at temperatures in the range of $1800°$ F. The presence of sulfur in most hot gases causes very rapid deterioration of turbine blading, especially if alkaline metals are also present.

A promising process for high temperature sulfur removal appears to be the reaction of sulfur with calcium oxide which may occur in a fluid bed gasifier. Objectives of this development work should be the following:

a. Minimum sulfur in effluents,

b. Minimum pressure drop,

c. Reliability (i.e., a transient breakdown which releases excessive sulfur into the effluent could possibly destroy the turbine), and

d. Optimum recycling of the sulfur absorber material.

7. *Sulfur Removal at Low Temperatures*

Where the product gas from the gasifier is to be converted to high BTU gas, the gas must be cooled and the hydrogen sulfide must be reduced to below 1 ppm before the methanation step. This can be accomplished in a scrubbing system, of the kind used presently in coke-oven scrubbing systems, but the pressure in the latter case is atmospheric and the degree of clean-up required is less stringent. At the high pressures in coal gasification systems, corrosion (particularly stress corrosion) will be enhanced because of the higher concentration of corroding species encountered in the scrubbind liquid. The higher concentrations result from the higher pressures involved.

The objective of development work in this area is to evaluate the corrosion resistance of presently available and newly developed materials in the expected environments.

8. Heat Exchangers

Optimizing the cycle efficiency involves waste heat boilers and other heat exchanging processes at the combustion interface. Problems of impingement damage to tubes (if hot gas is on the shell side), inlet erosion (if the hot environment is on the tube side), steam leakage across the barrier, crevice attack, flow-induced vibration, hot spots produced by deposits, etc., will reduce the reliability of heat exchanger operation. Objectives of research in this area will involve identification of ways of preventing such effects from causing failure.

Such units should also be designed for ready tube plugging or tube removal.

MATERIALS EVALUATION AND MODIFICATION

The purpose of engineering materials evaluation and modification is to tailor and adapt specific materials for the components in coal gasification units. Such materials must be readily fabricable, reliable over the full range of conditions expected in the specific component, and readily available.

The evaluation of these materials will depend substantially upon well controlled tests in a materials test facility as well as in operating pilot plants and in demonstration units. Supporting laboratory work will also be required.

The materials selected will be within the framework of existing materials systems. In some cases, commercial materials can be used with no modification; in other cases, existing materials must be modified, involving alternations in composition and structure or the development of composite materials such as weld-deposited hard facing on substrates of high toughness. Nonetheless, the evaluation of existing materials for these new and aggressive environments will involve an extensive testing program.

The necessity for materials research is defined by circumstances requiring materials for which there is no industrial process from which reliable materials systems can be transferred directly. Also, the relatively large size alone of coal gasification systems provides an additional set of uncertainties.

Further problems have been identified by applying existing knowledge from laboratory experiments which suggest imminent failures at expected combinations of operating conditions. There is also simply the intuitive expectation that the higher temperatures, erosive conditions, and well known noxious species such as sulfur will inevitably cause accelerated degradation.

The following materials systems are promising for the intended application, and early efforts should be directed toward their evaluation:

• Pressure Vessel Shells. ASME code-approved carbon and low alloy steels with possible improvements of the interior surface by procedures such as weld depositing of stainless steel.

• Ceramic Liners. High density alumina for severe environment ceramic surfaces and porous products for interior insulation. Possible improvements in erosion and wear resistance by chromia additions.

• Wearing Surfaces of Valves and Seals. Cobalt-base alloy, weld-deposited surfaces and tungsten-carbide inserts.

• Catalysts. Raney nickel with possible improvements in methods for fabrication and modification for extending the useful life.

Despite the variety of processes under development for gasifying coal, the "materials problems" are common. Since certain economies in the materials development work are possible because of this commonality of problems, they may be approached with a thoroughness which will raise the confidence of expected resolution.

The specific areas of materials engineering development considered to be common are the following:

1. Erosion and Wear Resistance of Metals and Ceramic Materials at High Temperatures in Complex Environments

High velocity gases containing entrained particles are expected in the gasifiers, transfer lines, and turbines. Maximum velocities in the range of 50-100 50-100 feet per second are expected in the first two. In the turbine, the velocities are expected to be 300-800 feet per second. In fact, the particulates must be reduced to a negligible amount and size for the turbines to operate. There is already evidence that flow eddies at joints and valves produce substantially increased erosion rates. There is virtually no information which relates the erosion rates to the oxidizing-reducing conditions and to the chemistry of the reaction product scale. This is significant because the mechanical properties of the scale vary greatly depending upon even subtle changes in the chemistry of the alloy.

In addition to the particulates, larte mass flows of hot solids are also transported at temperatures to $1800°$ F. These include the transfer of dolomite, hot ash, and slag.

99

2. **Erosion and Wear Resistance at Low Temperatures**

Large quantities of solids must be transported at room-to-moderate temperatures. A particularly important and unique transport process involves feeding the coal into the high pressure gasifiers. A successful process has yet to be found for continuous feeding; current batch feeding involves frequent opening and closing of valves which, as a result, are expected to sustain aggravated wear. Here, again, the transfer lines will be subject to wear although not at the high temperatures. However, the possible presence of moisture and associated leached chemicals from the coal will accelerate the wear.

3. **Corrosion Resistance of Metals and Ceramics in Hot and Complex Oxidizing-Reducing Gases**

Without the aggravation of abrasive processes, the gaseous environments in coal gasification are still corrosive. Such species as sulfur, hydrocarbon gases, hydrogen, alkali metals, chlorine, vanadium, and oxygen in various combinations are already well known to be very corrosive. Experience in the turbine industry has shown that small amounts of sulfur in combination with alkali metals attack the blades very rapidly. Phenomena known as carburization, metal dusting, sulfication, halidation, oxidation, and catastrophic oxidation studies in the laboratory, have already been seen in similar commercial equipment in other industries and have established sufficient precedent for expecting that they may occur in coal gasification. Ceramic materials experience similar processes of chemical degradation but also suffer from the existence of porosity, cracks produced by thermal transients, and the presence of impurities. These factors aggravate premature deterioration. Gaseous attack of metals is further aggravated by slight thermal cycling and by cycling changes in the oxidizing potential.

While much of the metal surface will be protected by ceramics, present experience suggests that ceramic materials may occasionally break and expose the vessel wall directly to the aggressive environments. In such cases, the metals should have such sufficient resistance to attack that the integrity of the large vessel is not impaired before corrective action could be taken.

4. **Corrosion Resistance of Metals and Ceramics in Water and Steam Solutions of Concentrated Acids**

In the use of scrubbing systems for removing impurity gases such as carbon dioxide and hydrogen sulfide, solutions are produced which are extremely corrosive and also capable of causing stress corrosion cracking. The use of liners here will mitigate some of this concern but these systems have yet to be developed.

5. **Mechanical Stability of Metals and Ceramics with Respect to Fracture Toughness, Creep, Fatigue, Stress Corrosion Cracking, Hydrogen Damage, and Corrosion Fatigue**

The mechanical strength of materials is reduced when exposed to noxious environments. The environments prevailing in coal conversion are generally recognized to be particularly aggressive in this regard. In many cases, the synergistic effects of environments combined with applied stresses reduce the load carrying capacity of metals by as much as a factor of ten. These phenomena are usually alleviated by lowering stresses or protecting the surfaces.

The historical pattern of such environmental-mechanical degradation suggests that new deterioration processes may operate in coal gasification systems and will need to be defined.

6. **Integrity of Welded Structures**

The very large sizes of equipment planned for use virtually assure us that much of the equipment will be fabricated in the field. Such field fabrication raises substantial questions of weld quality. Welded joints in existing technology--even where the vessels are fabricated within controlled facilities--are still regions of concern.

Such considerations as heat treatment, impurities, inspection, automatic processes, and welding speed will need special attention. We anticipate that special attention will also have to be given to the final properties of welds under the mechanical circumstances of paragraph "5" and the chemical circumstances of paragraph "3" above.

7. **Characterization of the Properties of Coal**

The physical and chemical properties of coal vary greatly. These properties affect the process of gasification as well as the material of construction. Further, the energy required for coal preparation may be affected by prior chemical treatments. These variables in the properties of coal also need defining so that engineering and laboratory testing can cover with assurance the breadth of conditions of large operating plants.

8. **Effectiveness, Fabrication and Lifetime of Catalysts**

Catalysts for methanation and water gas shift are central elements for producing high BTU gas from coal. The critical catalysis step is methanation. Catalysts are usually made from a porous nickel product called "Raney Nickel." These catalysts become depleted or ineffective after a relatively short time (several months). This degradation appears to be related both to poisoning effects of environmental species such as sulfur as well as to localized changes in the surface structure. The replacement and rejuvenation of these catalysts is costly. Substantial advances both in resistance to degradation as well as more economical rejuvenation or replacement are required.

9. **Optimization of Physical and Mechanical Properties of Ceramics**

The ideal ceramic materials for resisting chemical and mechanical degradation have high density. These materials are also thermally conductive; however, a very low thermal conductivity is required for insulating purposes. Optimizing these opposing trends is highly desirable--for example, by multilayer design and construction.

COAL LIQUEFACTION

PROSPECTS

To construct a plant of 100,000 barrels-per-day capacity of heavy fuel oil would require an investment of about $550 million (1974 dollars). Such a plant would produce a liquid fuel which would sell for about $1.15 per million BTUs, assuming petroleum industry economics and Pitt seam bituminous coal at $8.00 per ton. A somewhat lower cost of about 80-90 cents per million BTUs can be estimated on the basis of utility industry economics.

Several large oil companies are presently working on coal liquefaction processes, and at least one has formulated a schedule for placing one plant of 50,000 barrels-per-day capacity on stream by the early 1980s. The Office of Coal Research also has a program to begin the engineering work on a large-scale demonstration plant in early 1975. It is expected to be on stream by 1985.

CURRENT PROGRAM AREAS REQUIRING ASSESSMENT

1. <u>Temper Embrittlement of Chrome-Molybdenum Reactor Steels</u>

Some research is now supported by the Metals Properties Council and the American Petroleum Institute. Recently, a major oil company discovered cracks in high pressure reactors constructed from 2½ chrome-1 moly steel. Examination of the failed steel revealed that its 40 ft/lb Charpy transition temperature had shifted from about 32° F to about 300° F, after several years of service in the range of 700-800° F. This experience resulted in the American Petroleum Institute advising industry to operate 2¼ chrome-1 moly and 3 chrome-1 moly reactors at 20% of design stress when below 250° F during start-up and shut-down operations to reduce the possibility of catastrophic brittle failure. This possibility is considered to be a matter of prime concern since reactor materials and operating conditions for coal liquefaction are very similar to the conditions which have led to temper embrittlement in refinery reactors.

2. <u>Non-Destructive Materials Testing and Inspection Techniques</u>

Research in this area is being conducted by a variety of organizations, including the nuclear industry. A particular need is for automatic monitoring systems for continuous measurement of equipment wall thickness where erosive wear is significant.

3. <u>Effects of Hydrogen on Chrome-Moly Reactor Steels</u>

Some research is presently being supported by the Metals Properties Council and the American Petroleum Institute. During operation, reactor steels absorb large quantities of hydrogen. This hydrogen can have major effects on the properties of the reactor steel both at high temperature operating conditions and at low temperature shut-down conditions. At high temperature, the hydrogen can react with metal carbides causing severe damage. This type of damage is successfully avoided by employing alloys containing carbides stable within the reactor operating conditions of temperature and hydrogen partial pressure. Also, at high temperature, hydrogen may degrade the mechanical properties of the steel employed. Less is known about this effect. As a result, the Metals Properties Council is sponsoring work on this potential problem at the Southwest Research Institute. At low temperatures, the atomic hydrogen absorbed in the reactor steel at high temperature has reduced solubility resulting in the formation of molecular hydrogen. This molecular hydrogen can lead to hydrogen cracking of reactor steels. Work on this problem has been sponsored recently by the American Petroleum Institute at Battelle Memorial Institute (Columbus, Ohio).

4. <u>Engineering Parameters for Two and Three Phase Flow</u>

A cooperative research program is being conducted under the auspices of the American Institute of Chemical Engineers. Many materials problems can be avoided by adequate process and mechanical designs; with coal liquefaction, an improved understanding of engineering design parameters for multi-phase flow (solid-liquid-gas phases) should permit process designs which will avoid or reduce many materials problems.

5. <u>Sour Water Corrosion</u>

Some research is presently supported by the American Petroleum Institute. Large quantities of sour water containing hydrogen sulfide, ammonia, hydrocyanic acid, phenols, organic acids, chlorides, etc. will be produced by the coal liquefaction process. Equipment for handling these types of waters has suffered severe corrosion problems in the petroleum and steel industries. Sour water stripper overhead condensers, reflux pumps, and certain piping have been the principal problem areas, with austenitic stainless steel equipment failing in as little as one to two weeks.

AUTOMOTIVE GAS TURBINE ENGINES

PROSPECTS

Since automobiles consume about 40% of the nation's fossil fuel supplies, while miles-per-gallon efficiency has been decreasing, the adoption of ceramic gas turbines promises to effect important gains over current piston engine technology. If development is continued, it is possible that high temperature ceramic gas turbines for passenger cars could be in limited production by 1985, provided that the machine tool industry could be significantly expanded in parallel.

COMPARISON WITH CONVENTIONAL ENGINES

In assessing a potential energy impact of ceramic gas turbines, one must consider what efficiency can be expected over the years, and compare it with the specific fuel consumption of a conventional metal gas turbine of current technology (Figure 8).

per gallon over a 1974 piston engine of comparable performance. As the temperature of the ceramic engine is raised, the performance advantage will increase to approximately a factor of 2.3 over the 1974 engine. This potential advantage is illustrated in Figure 9.

PRODUCTION CONSIDERATIONS

Production changeover to ceramic gas turbines would require a capital investment of $250 million for each line capable of producing 500,000 engines per year. Two or three of these could be brought on stream each year starting in 1985. With an aggressive program, the plants installed in 1990 could produce engines with inlet temperatures of 3,000° F, with corresponding low fuel consumption. These assumptions lead to the estimates shown below:

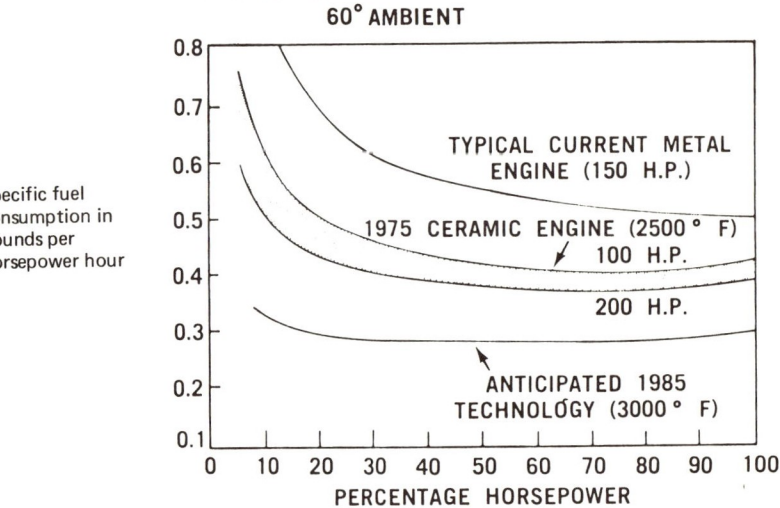

Figure 8. *Specific fuel consumption versus percentage horsepower for several turbine engines. There is no comparable unique curve for an internal combustion engine, but the upper line shown here can be regarded as a rough approximation for comparison purposes.*

A reasonable way to compare gas turbine engines with internal combustion engines is by means of a typical driving cycle comprising 25% city, 38% suburban and 37% at 50-miles-per-hour driving. For these test conditions, a 1975 ceramic engine illustrated by the central curve in Figure 8 would provide a factor of 1.3 advantage in miles

Table 13

Year	Cumulative Incremental Investment (Billion dollars)	Cumulative Oil Savings (Billion gallons)
1990	2.5	4.6
1995	5.5	31.5

In addition to the benefit of efficiency, the controlled combustion in turbines leads to low emission of NO_x and hydrocarbons so that it may be possible to meet the low stringent legislative standards without additional control devices.

Ceramic turbines will be made from silicon nitride and silicon carbide materials of which the basic constituents are in abundant supply. The turbines can operate on a variety of fuels.

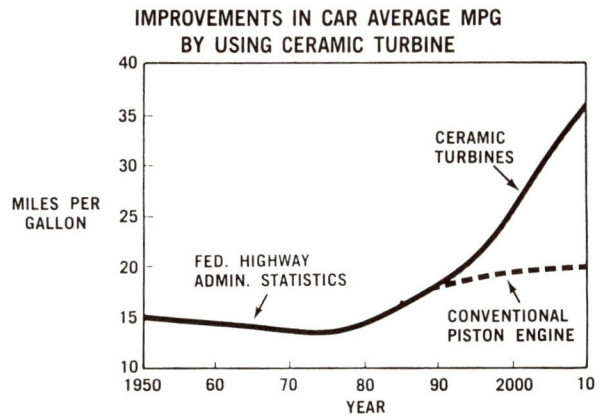

Figure 9. *A forecast of the average miles per gallon for all automobiles using conventional piston engines versus the result of starting an aggressive program to introduce high-temperature ceramic gas turbine engines in about 1990.*

CRITICAL ELEMENTS

BACKGROUND

The proposed energy program implies a pattern of substantial requirements for speciality materials. These requirements, superimposed on those routinely projected for American industry, are concentrated within the steel industry, and within that sector, on materials to tolerate high stresses and temperatures, severely corrosive and erosive environments, and various combinations of these.

Analysis of these requirements by the committee led to the identification of nine materials that warranted review as to adequacy of their supply to meet this expanded U.S. demand. Although precise quantitative assessments could not be made without exact, time-phased requirements figures (and these are not available), it was possible in a general way to establish relative degrees of the need for advance action to improve supply-demand balance to meet future capital requirements of the energy program. Specifically, the materials were evaluated against such criteria as:

- essentiality for the energy program,
- extent of reliance on imports,
- potential adequacy of substitutes, and
- vulnerability of imports to concerted control of price or flow.

On this rough basis, an approximate ranking of the materials in order of their "criticalness" was determined as follows:

1 - Manganese

2 - Chromium

3 - Fluorspar

4 - Nickel

5 - Cobalt

6 - Aluminum (including bauxite)

7 - Tungsten

8 - Platinum (including platinum-group metals)

9 - Copper

Emphasis was on the short-range future (1985). This policy excluded consideration of projects to develop new alloy systems or substitutes involving extensive redesign of systems.

Main attention was focused on specialty requirements of the steel industry. One surprise was the appearance of fluorspar as a rather seriously "critical" material-- attributable to its essentiality in both steel and aluminum production.

Because the energy program has been identified as a national goal, it was assumed that this program would have prior rights to available supplies of scarce materials. Accordingly, conservation actions would need to be addressed to non-energy uses of materials or to general improvement in supply-demand balance.

CRITICAL ELEMENTS

Given these considerations, the committee identified for the nine selected materials the following options in supply-demand correctives and "quick pay off" research and development.

MANGANESE

1. Possible Actions or Potential Solutions

    a. Look for additional higher quality domestic supplies and develop technology to use lower grade domestic supplies, but recognize that any significant contribution is more than a decade away.

    b. Develop an undersea supply source of manganese nodules.

    c. Reduce the use of manganese in steelmaking by:

    (1) Using basic oxygen processes which use slightly less manganese, and

    (2) Reducing the amount of manganese in carbon and low alloy steels by lowering sulfur and using some substitutes (e.g., nickel or molybdenum) for alloying elements.

    d. Look for additional high grade foreign sources.

    e. Stockpile.

2. Research and Development Options

    a. The use of manganese to reduce sulfur could be reduced (25%) at some small increase in production cost, by treatment of molten steel with calcium silicate. Minor commercial work of an empirical nature might be warranted.

Some empirical work could be done in the controlled rolling and cooling of steel to reduce grain size, enabling a small reduction (25%) in the use of manganese as an alloying addition to strengthen steel.

Preparatory research in the processing of undersea nodules could be accelerated.

CHROMIUM

1. Possible Actions or Potential Solutions

    a. Substitutions

    Currently many applications for stainless steel (e.g., stainless flatware, stainless sinks, automotive trim and wheel covers) could be switched to other materials with perhaps some economic and/or aesthetic sacrifice but at a saving of chromium for applications for which there is no substitute.

    b. Reduce Chromium Content in Current Application

    In many applications the chromium required on the surface for corrosion or oxidation resistance is put there by diffusion, cladding, or weld overlay techniques. Use of these techniques can conserve chromium, especially in heavy sections of what is now all high chromium alloy.

    Austenitic stainless steels contain chromium for corrosion resistance and austenite stability. Increased use could be made of other austenite stabilizers (nickel, manganese, and nitrogen), thereby reducing chromium content by about 10% (18% to 15%). In ferritic stainless steels where the carbon is combined as chromium carbide, a reduction of carbon from 0.05% to 0.02% would permit a reduction in chromium content of ½%.

    A substitute for the catalytic converter to control automotive emissions (for example, a stratified charge engine) could make available a 5-6% increase in the chromium supply.

    Development of lower chromium oxidation-resistant alloys employing aluminum and/or silicon could conserve chromium.

    Employment of boron as an additive can enable retention of hardenability in engineering steel with reduced content of chromium and other hardening additives.

    c. Expand Foreign Supplies

    Formulate a program to look for and develop a broader foreign supply base in potentially more stable countries.

    d. Expand Domestic Supplies

    (1) Look for additional domestic supplies, and

    (2) Develop processes to use lower grade ores.

    e. Stockpile

2. Research and Development Options

    a. Development of cladding and/or coating techniques more acceptable to the user, to put high chromium contents on the surfaces where they are needed, including the design-engineering-development to use such materials with reliability. The technology largely exists, but 1-2 investigators might be warranted to assist the users in the application of these techniques over the next ten years or until problems are resolved.

    b. Development of lower chromium stainless steels for less critical applications and the expanded use of those lower chromium alloys already available. Some of these alloys exist, but additional research at the level of 3-5 principal investigators would improve prospects of early use.

    c. Development of an alternative to the catalytic converter for automotive emission control.

    d. Development of oxidation and hot corrosion resistant alloys using aluminum and/or silicon in place of some or all of the chromium. Such a program could usefully engage the attention of 3-5 principal engineers.

FLUORSPAR

1. Possible Actions or Potential Solutions

    Because all uses tend to be consumptive, there is no potential secondary supply. Strategies for assuring sufficient supply to meet needs are limited to more extensive exploration, already supported by the Office of Minerals Exploration; substitution of other fluorine-containing raw materials for spar; conservation by reduction of non-energy-related uses of fluorine; and substituting other fluxes for fluorspar in the iron and steel industry.

    The commonest alternate fluorine-bearing mineral is phosphate rock, which is currently mined for fertilizer production. Contained fluorine is released as hydrofluosilicic acid, and there is strong pressure from environment-oriented bodies for its recovery. Currently two plants are recovering about 65,000 tons per year, most of which is used by the aluminum industry to make synthetic cryolite flux, aluminum fluoride, and hydrofluoric acid. Recovery from this source is expected to increase until all phosphate-processing plants are recovering fluorine; however, it is estimated by the Bureau of Mines that this resource will have been depleted before the end of the century. Because world fluorine reserves are considered to be adequate, imports and stockpiling to smooth out supply fluctuations represent the option that is most likely to be selected.

An electrolytic reduction plant using aluminum chloride is now under construction. Wider use of this process in new aluminum-plant construction would reduce the fluorine demand materially. Full conversion will reduce total fluorspar consumption by 10% and also reduce energy consumed in aluminum smelting by one third.

Shortage of fluorspar is most likely to occur in the cheaper metallurgical grade, acid-grade spar being readily produced by flotation concentration. The lower-grade lump product is, however, preferred by steelmakers because of both low price and higher recovery due to larger particle size. At present, an increasing proportion of spar consumed in steelmaking is the form of pelletized concentrates. There is potential here for releasing fluorspar to the aluminum, petroleum, uranium, and chemical industries if an adequate substitute can be found.

2. Research and Development Options

Current knowledge of other slag-flux systems is scant, especially for halide fluxes. Further study will be needed to determine substitution possibilities. One possible substitute, now being produced at a rate of about five times consumption, is calcium chloride, a byproduct of the Solvay soda process; at present the excess simply is dumped. The effect of calcium chloride on melting temperatures, chemical activities, and fluidity of basic-oxide slag systems warrants further study for both economic and environmental reasons. If feasible, it could eventually reduce fluorine demand by 40%, although the full effect is unlikely to be felt by 1990.

A possible route to reclamation of fluorine is to extract it from basic oxygen furnace slags, which contain between 5% and 10% calcium fluoride. The form in which it occurs is unknown. The information to be gained from a mineralized study of fluorspar-bearing basic slags could lead to a process for secondary fluorine recovery.

U.S. fluorspar imports are governed by transportation economics rather than politics; thus, over three-fourths of all imports, including nearly all metallurgical-grade imports, come from Mexico, a country with a stable, if independent political climate, in which transfer to local ownership of American-owned mining firms has already occurred. The possibility of price increases, however, remains. Research and development efforts to eliminate this dependence include the following:

a. Toward substitution for fluorspar as a steelmaking flux: A study of phase relations and chemical activity, and liquid-phase fluidity in metal-oxide/silicate/halide systems, especially those including calcium chloride. Suggested Level: 3-5 principal investigators over 3-5 years.

b. Toward secondary fluorine recovery from slag: Initial: a mineralogical study of basic oxygen furnace slags, in order to determine which fluorine-containing mineral phases are present, at various slag compositions and cooling rates. Follow-up: determination of leachability of fluorine from slag containing fluorine in such minerals. Suggested Level: initially, 1-2 principal investigators over 2-4 years; follow-up, same level.

NICKEL

1. Possible Actions or Potential Solutions

Nickel conservation can, of course, be sought wherever nickel is used. But it is most rewarding to look for conservation opportunities where the use is most extensive. Categories of conservation action are presented below in accordance with this principle.

a. Reduction of Nickel in Stainless Steel

The use of manganese and nitrogen to replace part of the nickel in austenitic stainless steels has been practiced in the seel industry since the mid-1950s. Some 30,000 tons of so-called AISI-200 series steels are produced annually, reducing nickel consumption in this use by 50%, with the further benefits of greater strength and a $40 reduction in price per ton. Some 45,000 tons of nickel are used to make stainless steel, upwards of perhaps 20,000 tons of nickel could be saved by maximum use of the 200 series. It should be noted that this action would result also in economizing use of chromium. It is also possible that some austenitic (i.e., nickel-bearing) stainless steels might be replaced by the less formable and weldable ferritic (non-nickel) stainless steels.

b. Reduction of Nickel in Plating

While nickel plating consumes a very large quantity of nickel, there are two opposite issues here. On the one hand, nickel plating over steel serves to conserve material against deterioration. On the other hand, much of this nickel is used as undercoat under hard chrome plating as a "cosmetic," particularly on automobiles, architectural trim, and consumer durables. If chrome plating is to be used, the undercoat is desirable; however, such trim could be replaced by aluminized plastic, tin-plate, or thin-rolled stainless cladding, etc.

 c. <u>Reduction of Nickel in Alloy Steels</u>

  Conservation of nickel in engineering alloy steels was effected to a large extent during the Korean War by the introduction of boron into the alloy. This treatment enabled reduction in alloy content by intensifying the hardening effect of the nickel, chromium, and molybdenum.

 d. <u>Other Possible Reductions</u>

  In cupronickel, used in boiler tubes, some specifications may still call for 70% copper, 30% nickel. An iron-modified 90% copper 10% nickel alloy is generally satisfactory, with the added advantage that it inhibits marine growth.

  Stainless steel cladding can sometimes replace solid stainless steel, where surface attack is a problem. Chromizing can also be considered.

  It should be noted that large experience was gained with nickel conservation measures, particularly during the Korean War. This experience is quite well documented. Specific details of opportunities for nickel conservation in particular applications could be sought out and employed if:

  (1) The necessity was made evident,

  (2) Motivation was provided,

  (3) The technology was technically proved, and

  (4) Cost factors were reasonable.

For example, the extensive use of stainless steel in hospital ware is open to question. Considerable substitution of glass for stainless steel in the dairy industry and in some chemical plants is a possibility.

 e. <u>Other Means to Balance Nickel Supply with Demand</u>

  Attention to the possibility of an increase in U.S. nickel production from domestic mines, or from other unexploited areas of the world, might warrant consideration. Large nickel deposits in northern Ontario, proved and developed by the International Nickel Company, could be developed if it became necessary and economically attractive to that company. The use of a commercial reserve stockpile to assure a sustained demand has been suggested. Over a longer term the significant fraction of nickel in manganese sea bed nodules could be exploited.

 2. Research and Development Options

  A few research tasks are here projected, on the basic assumption that there will be a need to divert a part of available nickel supplies to the energy program by 1985.

  a. Improve the usefulness of the AISI-200 series of low-nickel stainless steels. Prepare a handbook to characterize the full series of 200 steels, filling in gaps by test where necessary. Estimated time required: three years. Estimated cost: $500 thousand. Estimated contribution to nickel conservation: reduction by a further 10-30% in the 45,000 tons of nickel employed in current stainless steel. The basic control action could be applied immediately and intensified as information from this task became available in three years.

  b. Exploratory research to identify possible uses of the iron-aluminum system to replace some or all of both chromium and nickel; possibilities would include substitution for nichrome resistance wire as furnace heating elements, etc. The approach would take into account feasibility of producing low-carbon aluminimum iron in basic oxygen furnaces. Research directions would include testing of carbide formers in the system (molybdenum, titanium, etc.). Estimated time required: five years. Estimated cost: $2 million. Estimated contribution to nickel conservation: initially, 3-5% of nickel used, and substantial cost reduction in the product. Some application of the work could become available in one year, and progressive further results thereafter. The United Kingdom has developed an interesting austenitic alloy containing 1.0 carbon, 25 manganese, 10 aluminum, balance iron. Investigation and elaboration of this system could proceed at a more rapid rate, based on the British findings at modest cost ($1-3 million over 5 years).

  c. Further "fill-in" research to characterize the role of boron in hardenability of steel. Preparation of guidelines of boron and other less critical-elements substitution in place of nickel-containing engineering alloys. Preparation of handbook on hardenability. Estimated time: one year. Cost: $100 thousand. Estimated potential nickel savings: some 10-30% reduction in the 20,000 tons of nickel used in alloy steels might be attainable through this application, with effective guidance that this manual would provide. Savings could become effective immediately and progressively thereafter.

d. Research on fundamental understanding of low temperature toughness in steels, to enable use of lower-nickel alloys for these applications. Estimated time required: 5-10 years. Cost: $2 million. Potential savings of nickel: some 5-10% reduction in the 20,000 tons of nickel used in alloy steels as an ultimate objective, coupled with improved performance in low temperature performance of alloy steels.

e. Further investigation of very high (0.5 to 2.0%) nitrogen additions in steel. Potential savings of both nickel and chromium could ultimately result, but this would be a more speculative area in terms of possible results.

COBALT

1. Possible Actions or Potential Solutions

Because the tonnage consumption is comparatively small, cobalt supply could perhaps be assured by maintenance of a commercial stockpile of moderate proportions.

Nickel can be substituted for cobalt in most applications in general as well as in energy-related applications in particular. For example, nickel-base superalloys could be used instead of cobalt-base high temperature alloys for vessel shells and gas turbine components, although with some possible loss in performance. Likewise, certain nickel- or iron-base alloys can be substituted for cobalt-base alloys in hard-facing applications, but again with possible loss in performance. On the other hand, no completely satisfactory substitutes have been developed for cobalt in the production of bonded carbides for cutting tools and dies, although ceramic cutting tips are a partial substitute for cemented carbides in this use. An emerging possibility is the use of synthetic carbonadoes (polycrystalline diamonds) for ultra-hard tools and dies.

In an emergency, the U.S. could achieve temporary self-sufficiency by working its own limited reserves, estimated to contain about 28,000 tons of cobalt. Several mines and mills could be reactivated, notably the Calera Mining Company property at Cobalt, Idaho. This property had been developed in the early 1950s with U.S. Government support to establish the cobalt stockpile. In the longer term, the cobalt fraction of manganese sea bed nodules may prove significant.

2. Research and Development Options

It is judged that research and development efforts will not be useful in the context of the cobalt supply-demand situation within the time scale of this study. Minor economies could possibly be effected in the use of combined nickel and cobalt as a binder of carbides for machine tool cutting tips.

ALUMINUM

1. Possible Actions of Potential Solutions

Abundant alumina-making raw materials such as clay, alumite and anorthosite occur domestically, and could probably replace imported bauxite at a cost of 3-5 cents per pound of metal.

The use of aluminum reduction processes using less energy is to be encouraged.

Attention should be given to the recycling of aluminum scrap from municipal wastes. On the order of 3 million tons of aluminum annually finds its way into such repositories as sanitary landfill. Remelting of aluminum scrap requires only one twentieth of the electricity needed to produce an equivalent weight of virgin metal.

2. Research and Development Options

The present minor efforts to process clay and alumite to make alumina should be expanded. Pilot plants of about 25 tons per day capacities should be constructed at an early date to prove out extraction technologies, followed by construction of demonstration plants to establish cost data upon which the design of full scale plants may be predicted.

This program is expected to cost on the order of $10 million for pilot plant construction and operation plus $50-75 million for demonstration plants.

Target dates for demonstration plant construction should be 1976 or 1977, towards the goal that some full scale plants producing as much as one million tons of alumina per year might be in operation prior to 1985.

The prospects for success for such a program are excellent, based upon the extensive prior work by private industry and government. The current cooperatively financed mini-pilot plant operation in Boulder City, Nevada, looks promising and will be more so as start-up minor hitches are eliminated.

Similar comments apply to the Hazen Research pilot plant at Golden, Colorado.

TUNGSTEN

1. Possible Actions or Potential Solutions

A two-pronged approach to increase domestic production and to develop non-tungsten-containing substitutes is suggested. Although our reserves are quite limited, low grade ore resources are extensive both in the U.S., Canada, and certain other friendly countries. Technology should be developed to most economically recover tungsten from these low grade ores as a guard against restriction in supplies from outside sources. Maintenance of a substantial commercial stockpile of tungsten metal and concentrates would be desirable.

Substitutes are possible for some uses of tungsten. Titanium carbide and synthetic carbonadoes (pressed and sintered artificial diamonds produced domestically) could replace much of the tungsten carbides used in drilling and metal machining as well as other wear and erosion resistant applications. Ceramic cutting tips can replace carbides in many uses. Furthermore, molybdenum can be substituted in part for tungsten in some tool steel and superalloy applications, but almost always at the expense of properties. These substitutions in steel and superalloys are quite well developed.

2. Research and Development Options

Research and development could minimize the exposure to drastic changes in tungsten supply from external sources. The options include:

a. Carry out development work on alternate carbides, and on artificial diamond-type materials as substitutes for tungsten carbide to obtain quantitative data on the effective use of these alternate materials.

b. Develop an off-the-shelf process for recovery of tungsten from low grade ore resources. Intensify the search for further occurrences.

PLATINUM-GROUP METALS

1. Possible Actions or Potential Solutions

Although it might be possible for platinum producers to increase production fourfold within the next several years, this is not going to happen by 1975, when exhaust converters become mandatory. Foreign mines cannot be opened economically for the short period before converter recycling begins to ease. Further, the two major supplier countries, while widely different in political attitudes, present problems to foreign-policy planners: on the one hand, the U.S.S.R. tends to adjust its trading policy to suit its political goals; and on the other, domestic political pressure may be mobilized to embargo imports from South Africa as a form of protest of its government's internal human-relations policies.

Because of the limitations of domestic reserves, requirements will have to be met by conservation, substitution, and increased imports. To allow for lack of access to foreign supplies, the domestic stockpile will have to be increased considerably. Conservation measures could include:

a. Elimination of such nonessential or dissipative uses such as jewelry, consumable thermocouples, and decorative plating,

b. Increased recycling beyond the present level of 70%, and

c. Less reliance upon platinum in essential uses where possible.

2. Research and Development Options

Research which could benefit platinum-metal conservation would be to investigate the means by which the specific surface area of platinum catalysts can be increased, so as to reduce the quantity required for a given application, and how the high-specific-surface condition may be stabilized. Knowledge of the mechanisms by which catalysts are poisoned would lead to increased catalyst lifetimes and reduce net apparent consumption.

Substitution possibilities are dependent on successful research into catalysis mechanisms and the electronic structure and chemical properties of materials known to have catalytic activity. Materials of interest here include: vanadium oxide, hydrofluoric acid, activated carbon and alumina, tungsten carbide, nickel, and cobalt. Research programs would, initially, be small, involving only a few principal investigators. As new catalytic materials are discovered, they will have to be tested in service, which will take several years per reaction per material. Although much work can be done simultaneously, obtaining results will take many years and it is unlikely that results of new research in this field will have much effect on energy production within the next 10 or 15 years. In the meantime, replacement of platinum by materials of known catalytic activity may have to be encouraged, regardless of cost.

COPPER

1. Possible Actions or Potential Solutions

To meet the required increase from domestic reserves and resources requires several actions. First of these is a determination of what sulfur dioxide levels in smelter gases may be tolerable without imposing unwarranted capital and operating costs upon smelter operators. Second, restrictions imposed upon access to favorable exploration grounds could be reviewed.

2. Research and Development Options

Sustained research effort is needed to develop environmentally acceptable concentrating process technology. Company efforts in this direction might be unified under a joint corporate-federal program to arrive at a solution earlier than can be attained by the present divided efforts. Possibly $10-15 million should be directed to this end, with a target goal of 5 years to arrive at the design of a demonstration plant.

Research and development efforts on the double-alkali and citrate processes should also be accelerated in order to obtain a solution to sulfur dioxide emission from existing plants. Funding on the order of $10-15 million could provide for construction and operation of demonstration plants for both processes.

TABLES

A summation of critical elements, uses, and sources is given in Table 14. Non-energy uses of critical elements are listed in Table 15.

TABLE 14.   Critical Elements: Uses and Sources.

MATERIAL	UNIT	ANNUAL U.S. REQUIREMENT	ANNUAL U.S. PRODUCTION		EXTENT OF RECYCLING		IMPORTS		U.S. POSTURE[3]	
			Amount	%	Amount	%	Amount	%	Known Reserves	Known Resources
Aluminum	Short ton	5,588,311	4,122,451	74	945,727[2]	17[2]	794,485	14	13m	740-3700m
Bauxite	Long ton	15,375,000	1,812,000	12			11,438,000	75		
Chromium	Short ton	1,140,000	0	0	-	-	1,061,000[6]	94	No data	Insignificant
Cobalt	Ton	6,450	0	0	-	-	6,340	98	23,000 st	No data
Copper	Short ton	2,240,000	1,660,000	74	455,194	20	415,600	19	81,000,000	70-186m
Fluorspar	Short ton	1,352,149	250,000	19	-	-	711,000	53	3,000[14]	2,000-7,000[14]
Manganese	Short ton	2,331,459	39,628	2	-	-	2,291,831	99	No data	90-100 m
Nickel	Short ton	203,214,000[8]	15,739,000	8	35,926,000	18	125,364,000[9]	62	Small	5-14m
Platinum - group metals	Troy oz.	1,559,822	15,380	Neg.	255,641	16	1,836,349	118	1,000,000[10]	6-12m[10]
Tungsten	Short ton	7,054[12]	4,075[12]	58	-	-	2,870[12]	41	87,500[9]	175-380[9]

WORLD POSTURE[4]		U.S. IMPORTS FOR CONSUMPTION[7]		
Known Reserves	Known Resources	Country	Amount	%
12-15 b[5]	Unknown	Bauxite { Jamaica	6,958,000	45
		Surinam	2,923,000	19
		Dom. Rep.	910,000	6
		Haiti	617,000	4
%	%			
So. Africa 64	So. Africa 78	Chromite { USSR	226,000	45
Rhodesia 34	Rhodesia 21	[Cr$_2$O$_5$ { So. Africa	113,000	23
USSR 7		content] { Turkey	53,000	11
Finland 6		{ So. Rhodesia	43,000	9
		{ Philippines	43,000	9
		Ferro- { So. Africa	31,519	35
		chrome { Japan	11,865	13
		[chromium { So. Rhodesia	10,656	12
		content] { Sweden	7,921	9
		{ Norway	6,777	7
	m.lbs	%		
Cuba	2,312	23	Zaire	2,170	39
USA	1,684	17	Metal { Belg/Lux	1,500	26
Zaire	1,500	15	{ Finland	580	9
Caledonia	850	9	{ Zambia	480	8
Zambia	766	8	{ Norway	400	7
Australia	650	7	{ Canada	280	5
Canada	550	6			
	m.st	%			
S. America	80	23	Total un- { Canada	-	36
USA	76	22	manufactured { Peru	-	22
Africa	53	15	copper { Chile	-	14
USSR	39	11			
Europe	25	7			
	m.st	%			
Europe	62	33	Mexico	-	67
Africa	39	20	Spain and Italy	-	33
USA	25	13			
Mexico	25	13			
Asia	25	13			
%	%			
S. Africa 46	S. Africa 46	Gabon	473,142	29
USSR 46	USSR 39	Brazil	404,972	25
Gabon 3	USA 12	Zaire	278,595	17
Brazil 2	Canada 3	So. Africa	142,354	9
		Australia	52,587	5
	m.tons[9]	%			
Cuba	20	22	{ Canada	97,250	78
Canada	17	20	{ Norway	17,295	14
USA	16	17	Includes { UK	4,135	3
New Caledonia	9	10	scrap { So. Africa	2,741	2
Philippines	8	9	{ So. Rhodesia	1,801	1
%	%			
S. Africa 47	S. Africa 84	USSR	736,264	40
USSR 47	Rhodesia 13	UK	589,711	32
Canada 4	USA 3	So. Africa	237,697	13
Colombia 1		Japan	111,875	6
%[13]				
China 53				
Canada 12		Ore and { Canada	1,634	29
USSR 12		concentrate { Thailand	883	15
N. Korea 6		{ Peru	814	14
USA 6		{ Bolivia	780	14
S. Korea 3				

NOTES:

[1] Metal

[2] Full-industry estimate: 1,045,000, or 19 percent.

[3] NCMP data.

[4] Bu Mines Paper No. 820.

[5] High-grade refractory mainly in Guyana and Surinam.

[6] Chromite ore.

[7] Bu Mines 1972 Minerals Yearbook data.

[8] Includes scrap

[9] Metal

[10] Platinum only

[11] At 1971 prices

[12] Concentrate, tungsten content

[13] WO$_3$ only

[14] Fluorine

st = short ton
t. = thousand
m.= million

TABLE 15. Non-Energy Uses (1973) of Critical Elements.

Material	Percent	Use	
Aluminum	24	Construction	[12 Electrical]
	15	Transportation	
	3.5	Refractories	
Chromium	5-6	Automotive catalytic converters (1975)	
		Sulfur-removal equipment	⎫
		Sewage plants	⎬ Environmental control
		Acid waste treatment plants	⎪
		Mass transit	⎭
Cobalt	16.0	High-temperature, high-strength alloys (1500/9400)	
	15.4	Carbide-cutting and wear resistance applications (1450)	
	1.5	Hard-facing alloy rods (182)	
Copper	48	Nonelectrical	
Fluorspar	40 (1972)	Iron and Steel-making	
		Aluminum smelting	
	33-1/3	Fluorocarbon chemical and plastics industries	
Manganese	95	Refining and alloying of steel	
		Batteries	
		Chemicals	
		Catalysts	
		Medicinals	
			Product or industry uses:
Nickel	33-1/3	Stainless steel	Aircraft
	20	Electroplating	Military hardware
		Wrought steel alloys	Transportation systems
		Super alloys	Power plants
		Nonferrous alloys	Consumer durable goods
		Cast iron alloys	Industrial production equipment
		Magnet alloys	Hospital ware
		Chemical applications	Dairy industry equipment
			Chemical and petrochemical plant facilities
Platinum		Automotive catalytic converters	
Tungsten	40	Tool steel and high-temperature alloy addition	
	40	Tungsten carbide	
	11	Lamp filaments and other electrical uses	
	9	Other	

BATTERIES

PROSPECTS

Development programs are underway to exploit the advantages offered by batteries for electric vehicle propulsion and bulk energy storage. However, the projected impact of these programs on the U.S. energy program in the decade 1975-1985 is probably minimal.

Batteries for load leveling will play only a minor role in the next decade; in view of the status of the battery development programs underway (see below) and the requirement for a ten-year life, it is probable that only prototype systems will be installed and undergoing evaluation in 1985.

With respect to vehicular batteries, the lead-acid battery has seen some technical improvements and has increased its penetration in applications such as shop mules, small specialty vehicles serving airports and short-haul urban delivery routes, etc. Battery-powered personal vehicles are unlikely to accomplish more than a modest penetration of the transportation market within the next decade during which the lead-acid battery will be the only fully tested, widely available system. Development, technically and economically, of feasible advanced batteries is expected only toward the end of this period.

ADVANTAGES

For load leveling applications, batteries offer the advantages of:

1. Short lead time for installation,

2. Distributed siting, enabling economies in transmission lines,

3. System stabilization,

4. Short-time expansion in system capacity, and

5. Rapid emergency capability.

In addition to load leveling, batteries can be utilized in transportation. As a further contribution to load leveling, battery-powered vehicles may be recharged during off-peak hours. The overall energy efficiency of a central generator-battery vehicle system exceeds that of the internal combustion engine by up to half, and the replacement of scattered small engines by a central prime energy source permits the more efficient removal of pollutants from combustion products prior to discharge to the atmosphere.

DEFICIENCIES

These advantages, however, cannot yet be realized because commercially available batteries do not meet the technical and economic criteria for vehicular use or for energy storage in utility systems. The stringent technical and economic objectives established for the desired applications are shown in Table 16. A comparison with today's batteries may be made by reference to Table 17.

TABLE 16. Technical Objectives

	Load Leveling	Vehicular Application
Cost ($/kwh)	20	25
Efficiency (%)	75	50-70
Life (years)	10	3-5
Life (cycles)	2500	300-500

Under today's battery technology, secondary (rechargeable) batteries include lead-acid, nickel-cadmium, nickel-iron, and silver-zinc. Nickel-cadmium and silver-zinc batteries, while having a number of desirable features, are ruled out of consideration for the present applications by virtue of their high cost.

LEAD-ACID TECHNOLOGY

Today's battery technology for automotive and industrial applications centers on lead-acid batteries. On the basis of performance for cost, this versatile electrochemical system dominates the marketplace. Three types of interest are compared with the above-mentioned batteries in Table 17. It should be noted that a wide range of manufacturing capability is represented in just this small sample from the many types of lead-acid batteries available. Golf car batteries are manufactured with the equipment and methods used for automotive batteries, i.e., on a mass production basis. On the other hand, motive power and other industrial lead-acid batteries are "hand made" in small lots. The energy density and cycle life values cited illustrate the trade-off range available in the design of lead-acid batteries. Long-life batteries can be made at the cost of additional weight.

NICKEL-IRON TECHNOLOGY

The nickel-iron battery was developed by Thomas Edison in 1901. Until well into the 1920s it was the most prominent type of secondary battery in industrial use. In more recent decades it has been displaced almost entirely from the market. The design concepts originated by Edison are still in general use. These involve packing active materials (mainly iron hydroxide and nickel hydroxide) into pockets or tubes formed of perforated nickel-plated steel. The resulting battery plate is extremely rugged and batteries giving 10-20 years of satisfactory service in traction or railroad car lighting use are common. The "conventional" nickel-iron battery suffers from such shortcomings as high cost and inefficiency during charging with resulting heavy evolution of gas and consequent need for frequent water additions.

DEVELOPMENT EFFORTS

Efforts to develop batteries with enhanced performance and low cost usually focus on the active materials which participate in the electrochemical reactions. The choice is limited, and generalized criteria for selection are low cost, high reactivity, and low equivalent weight. A low equivalent weight (weight per electrical equivalent) is advantageous for reasons of cost. Those cells shown in Table 18 are the focus of current major development work directed toward bulk energy storage for load leveling.

MATERIALS PROBLEMS IN ADVANCED BATTERIES

1. **Stability of Positive Electrode Support Current Collector**

    This is a universal problem because the positive electrode sees a high (oxidizing) potential, significantly higher during charge. Elevated temperatures and solid oxide electrolytes add to the severity. The need is for oxidation-resistant electronic conductors with only average mechanical properties.

2. **Insulators and Separators**

    For use in feed-through and other insulator requirements, impervious ceramic materials resistant to the electrolyte in question are required. The separator requirement is more complex; a porous body which will absorb electrolyte is needed. Resistance to attack by electrolyte, negative and positive active materials is essential.

3. **Materials Costs**

    Materials costs are of such significance as to demand identification as a

TABLE 17. Comparison of Today's Secondary Batteries.

Battery Type	Cost[a] ($/kwh)	Specific Energy Density by[b]		Life[c] (cycles)
		Specific Weight (Wh/kg)	Specific Volume (kwh/ft^3)	
Silver-Zinc	900	120	8.8	100/300
Nickel Cadmium	600	40	3.6	300/2000
Nickel-Iron	400	33	1.4	3000
Lead-Acid				
Motive power	50	22	2.6	1500/2000
Golf car	25	35	2.2	400
Elec. Vehicle	(d)	35	2.8	500/1000

NOTES:
[a] Cost to user.
[b] The values shown are for the 6-hour rate; values for the 2-hour rate are 10% to 35% lower, depending on the specific battery system.
[c] Cycle life depends on a number of factors, including depth of discharge, rate of charge and discharge, temperature, and amount of overcharge. Range shown is from most severe to modest duty.
[d] New product, price not available.

TABLE 18. Electrochemical Cells Under Development For Bulk Energy Storage.

Electrochemical Couple		Electrolyte	Temperature
Lithium	Sulfur	$LiCl \cdot LiI \cdot KI$ [a]	400-450°C
Sodium	Sulfur	$Na_2O \cdot 11Al_2O_3$ [b]	300-350°C
Zinc	Chlorine	$ZnCl_2$ (aq)	Room
Iron	Ferric Chlorine	$FeCl_2$ (aq)	Room

NOTES: [a] Molten salt mixture
[b] Solid ceramic

possible bottleneck. In present-day battery systems, materials costs represent between one-third and one-half of total manufacturing cost. By way of example, assume:

    a. A cost to the user of $24/kWh and a manufacturing cost which is 60% of this or $15/kWh; and

    b. Materials costs of 50% of manufacturing cost.

The allowable average total costs for materials is 15 cents per pound if the energy density is 20 Wh/lb, 30 cents per pound if 40 Wh/lb, and 75 cents per pound if 100 Wh/lb. Clearly this puts a stringent limitation on materials which can be used and will serve to put materials development programs into sharp focus. Present automobile batteries cost about 30 cents to $1.00 per pound.

The successful development of any one of the classes of batteries could have a profound impact on achieving energy independence and on improving the quality of the environment. Materials are a particularly important consideration in battery development where materials technology is at the root of solving the cost problem. In addition to availability and cost, most other materials problems involve insulators, most notably separators, and positive electrode current collectors and structure.

Presently, about 30% of the $11 million annual battery research and development effort is devoted to materials research. In fact, if one includes cell testing and post-mortem cell examination as a materials characterization program, the materials effort is a significantly larger proportion of the overall effort. Doubling of the materials work can easily be justified to better assure the success of the efforts and to reduce the research and development time required. It is particularly important at this time to expand materials-related efforts before decisions are made to undertake large-scale development of the most promising battery systems.

FLYWHEELS

ROTOR PROBLEMS

Much experience has been gained in the use of metal flywheels for various purposes, and advanced designs and uses are constantly being pursued. A concept currently receiving great attention is the application of fiber-composite materials in flywheel systems for various applications, particularly for vehicle propulsion and large-scale energy storage on electric utility systems. The simplest configuration for such a flywheel is a continuous-filament-wound rotor. This design has been tested in the laboratory, but rapid delamination of the rotor occurred under the influence of the radial stresses imposed by the centrifugal force field. These results made it clear that both the circumferential and radial stresses imposed on a flywheel rotor must be accommodated by a combination of design and material development.

TWO NEW DESIGN CONCEPTS AND ASSOCIATED MATERIALS PROBLEMS

Various new design concepts for fiber-composite flywheels have recently been proposed; two particular concepts are representative of these. The first configuration would eliminate circumferential stress problems by allowing the centrifugal forces to be handled by a multitude of thin rods emanating radially from the hub. Stress problems at the hub could be minimized by designing the hub so that a single rod passing through the hub, but not through the axis of the flywheel, forms two "bristles" of the flywheel "brush." For this flywheel configuration, the important materials problems to be solved would seem to be:

- the consistent fabrication of composite rods having uniform and accurately controlled properties,
- the characterization of deterioration processes under stress and fatigue conditions imposed by the flywheel duty cycle,
- the development of materials specifications, fabrication techniques, and flywheel designs to accommodate the capabilities of these materials,
- the development of new materials allowing the fabrication of flywheels having higher energy densities and longer lifetimes, and
- the production of such fiber-composite materials in quantities sufficient to satisfy the large potential markets.

The second representative flywheel configuration presents problems which may be more difficult to solve, but which offers a theoretical advantage in its energy storage capabilities. This concept attempts to accommodate the circumferential stresses in concentric filament-wound fiber-composit rings of a thickness which is small as compared to the overall flywheel radius. The use of such thin rings would minimize both the radial stresses within each ring and the transfer of these stresses from ring to ring. Some method of attaching adjacent rings, without introducing stress concentrations, is necessary. The most promising concept for this purpose is the use of an elastomeric material between adjacent rings. Such a material might be a composite material having a modulus of elasticity which is lower than that of the main rings. This elastomeric spacing ring may be designed to allow for the most appropriate relative displacement of the adjacent main rings to optimize the stress field.

Materials problems requiring research include:

- the winding of the rotor rings under proper filament tension and other conditions,
- the assurance of uniform and accurately controlled material properties,
- the characterization of design requirements and operating conditions to optimize the trade-off of performance and lifetime,
- the selection and development of proper concepts and techniques for producing and utilizing the elastomeric spacing ring material, and
- the proper materials design for handling the various transient conditions to be experienced in the duty cycle imposed by each application.

The requirements technically are very similar to those imposed by aerospace applications, and it is anticipated that a large part of the technology developed for that application can be applied.

PROJECTIONS

It is to be expected that fiber-composite flywheel technology will not have a major impact on the electric utility or vehicle industries before 1985. However, it is quite feasible to expect that a demonstration of the applicability of this technology to these industries can be accomplished by then. A recent projection indicates that the successful implementation of this technology in the electric utility and vehicle industries would require a level of fiber-composite materials production which is about four times the present level.

Projections of the economic and technical performance of fiber-composite flywheels are promising enough to warrant close attention, even when the projected rotor material is considered to be "E glass" (alumino borosilicate). This implies that the applicability of graphite, fused silica, and other advanced fibers might be even more promising, depending on the cost of producing flywheels composed of these materials. Not only are the material production and flywheel fabrication processes critical, but the handling problems occurring during transportation and installations will be important to the economics of the concept.

The development of fiber-composite materials having increased ratios of tensile strength-to-density can be directly beneficial to the technology of fiber-composite flywheels. Equally essential, however, are the accurate control of the material properties, the fabricability of these materials into the necessary flywheel structures, and, of course, the cost of producing flywheels from these materials. A function which can be of immediate benefit to this technology is the greater direct involvement of fiber-composite materials experts in the programs aimed at the development of such flywheels for specific applications in the electric utility and vehicle industries. It is clear that specific development programs must be organized and implemented.

REFERENCES

METHODOLOGY

Brookhaven National Laboratory projection data, kindly furnished in advance of publication, in a format similar to that of the Joint Committee on Atomic Energy.

Bechtel Corporation flow diagram.

COAL GASIFICATION

MIT Technology Review, May 1974, p. 45.

Perry, Harry, "Coal Conversion Technology," *Chemical Engineering*, July 22, 1974, pp. 88-102.

Task Force on Energy, "U.S. Energy Prospects, an Engineer's Viewpoint," National Academy of Engineering, 1974.

"The Materials/Design Interface in Coal Conversion Technology," Ohio State Conference on Materials Problems and Research Opportunities in Coal Conversion, April 1974.

"The Nuclear Industry 1973," published by The U.S. Atomic Energy Commission, WASH 1174-73, p. 19.

"Nuclear Power 1973-2000," published by The U.S. Atomic Energy Commission, WASH 1139-72.

1971 Annual Report to Congress by the U.S. Atomic Energy Commission.

1972 Annual Report to Congress by the U.S. Atomic Energy Commission.

Nucleonics Week, July 25, 1974.

"Projected Independence: An Economic Evaluation," MIT Energy Laboratory Policy Study Group, March 1974.

HIGH TEMPERATURE GAS TURBINES

The 1970 National Power Survey, Federal Power Commission, Part 1, USGPO, Washington, D.C., December 1971, p. I-3-2.

Bratton, R. J. and A. N. Holden, "Energy Conservation Through Ceramic Gas Turbine Power Generation," Government Energy Committee presentation, Advanced Research Projects Agency, March 5, 1974.

NUCLEAR

Westinghouse Steam Generation Symposium, Pittsburgh, Pa., March 1973.

Bush, S.H., "Structural Materials for Nuclear Power Plants," ASTM Gillette Memorial Lecture, 1974.

Bush, S.H. and R. L. Dillon, *Stress Corrosion in Nuclear Systems*, USAEC Advisory Committee on Reactor Safeguards (Proceedings of 2nd International Conference on Stress Corrosion, Unieux-Firmany, France), June 1973.

Gross, J.H., "Materials Consideration," PVRC Interpretive Report on PV Research, Section 2, WRC Bulletin No. 101, November 1964.

EPRI Private Communication on Monitoring of Stress Corrosion.

PVRE *ad hoc* Group on Toughness Requirements, "PVRE Recommendations on Toughness Requirements for Ferritic Materials," WRC Bulletin No. 175, August 1972.

"Second Institute Conference on Periodic Inspection," Institute of Mechanical Engineering, London, 1973.

"Flow and Fracture," ASTM, Special Technical Publication 479, 1973.

CONSERVATION

Patterns of Energy Consumption in the United States, Stanford Research Institute, Office of Science and Technology, Washington, D.C., January 1972.

The Potential for Energy Conservation, Office of Emergency Preparedness, Washington, D.C., October 1972.

Energy Conservation in the U.S.: Short-Term Potential 1974-1978, National Petroleum Council, March 1974.

Lincoln, G.A., "Energy Conservation," *Science*, Vol. 180, No. 155, 13 April 1973.

Lansberg, H.H., "Low-Cost, Abundant Energy: Paradise Lost?", *Science*, Vol. 184, No. 247, 19 April 1974.

Berg, C.A., "Conservation in Industry," *Science*, Ibid., p. 264.

Cytopoulos, E.P., Lazarias, L.J., and T.F. Widmer, *Potential Fuel Effectiveness in Industry*, Ford Foundation Energy Policy Project (in press).

Myers, J., *Energy Consumption in Major U.S. Manufacturing Industries*, The Conference Board, New York, N.Y., (in press).

Kellogg, H.H. and J. Tien, "Energy Consideration in Metal Production, Selection and Utilization," presented at AIME, Chicago, Ill., October 1973.

Kellogg, H.H., "Technology for Materials Supply," Appendix VI, First Report, Committee on Mineral Resources and the Environment, National Academy of Sciences, (to be released in 1974).

"Potential for Effective Use of Fuel in Industry," Thermo Electron Corp., Report No. TE-5357-71-74, The Ford Foundation Energy Policy Project, April 1974.

Sullivan, P.M., Stanczyk, M.H. and M.J. Spendlove, "Resource Recovery from Raw Urban Refuse," U.S. Department of the Interior, Bureau of Mines, Report of Investigation 7760, 1973.

Anderson, Larry L., "Energy Potential From Organic Wastes: A Review of the Quantities and Sources," U.S. Department of the Interior, Bureau of Mines, Information Circular 8549, 1972.

Appell, H. R., Fu, Y.C., Friedman, S., Yavorsky, P.M. and I. Wender, "Converting Organic Wastes to Oil. A Replenishable Energy Source," U.S. Department of the Interior, Bureau of Mines, Report of Investigation 7560, 1971.

Sullivan, P.M. and M.H. Stanczyk, "Economics of Recycling Metals and Minerals from Urban Refuse," U.S. Department of the Interior, Bureau of Mines, Technical Progress Report 33, April 1971.

"Conversion of Municipal and Industrial Refuse into Useful Materials by Pyrolysis," U.S. Department of the Interior, Bureau of Mines, Report of Investigation 7428, 1970.

GEOTHERMAL

Bowen, Richard G. and Edward A. Groh, "Geothermal--Earth's Primordial Energy," *Technology Review*, October-November, 1971, p. 46.

Robson, Geoffrey R., "Geothermal Electricity Production," *Science*, Vol. 184, No. 371, 19 April 1974.

Thomsen, D.E., "Power from the Salton Trough," *Science News*, Vol. 106, No. 28, 13 July 1974.

Lorensen, Lyman E., "Materials Screening Program for the LLL Geothermal Project," Lawrence Livermore Laboratory, Livermore, California, UCID-16513, May 29, 1974.

Walter, R.A., Stewart, D.H. and P.N. LaMori, "Evaluation of Small Power Plant Systems for Use with Geothermal Reservoirs," Battelle Pacific Northwest Laboratories, BN-SA-359, August 30, 1974.

SOLAR

Solar Heating and Cooling Demonstration Act, HR10952 et al., Hearings of Subcommittee on Energy of House Committee on Science and Astronautics, 93rd Congress, No. 24, 13-15 November 1973.

Solar Energy Research, A Multidisciplinary Approach, Staff Report, House Committee on Science and Astronautics, 92nd Congress, 2nd Session, Series Z, December 1972.

"Solar Heating and Cooling of Buildings," Westinghouse Corp., NSF-RA-N-74-023A, May 1974.

FUEL CELLS

Bacon, F.T., "The High Pressure Hydrogen/ Oxygen Fuel Cell," presented at the Atlantic City, N.J. meeting of the ACS, Sept. 1959.

Bockris, J. O'M., *Electrochemistry of Cleaner Environments*, Plenum Press, New York and London, 1972, p. 21.

Eklund, Lueckel and Law, "Fuel Cells for Dispersed Power Generation," IEEE paper, April 1972.

Grove, W.R., *Philosophical Magazine*, January 1839.

King, J.M. and S.H. Folstad, "Electricity for Remote and Developing Areas via Fuel Cell Powerplants," Paper Presented at IECEC Meeting in Philadelphia, August 1973.

O'Sullivan, J.B., "Historical Review of Fuel Cell Technology," Proceedings of the 25th Power Sources Conference, May 1972.

ISOTOPIC SEPARATION OF URANIUM$_{235}$

Energy Data from Edison Electrical Institute, Electrical Output Table, Volume 41, No. 26, June 30, 1973.

"Nuclear Power Growth 1974-2000, USAEC Report WASH-1139 (74), February 1974.

New Enrichment Plan Scheduling, USAEC, ORO-735, November 1973.

Uranium Enrichment by Gas Centrifuge, D.G. Avery and E. Davis, Crane, Russak and Company, Inc., New York 1973.

"Uranium Enrichment Laser Methods Nearing Full Scale Test," *Science*, Vol. 185, No. 4151, August 16, 1974.